D1029702

Results and Problems in Cell Differentiation

Series Editors:
W. Hennig, L. Nover, U. Scheer

34

Springer

Berlin
Heidelberg
New York
Barcelona
Hong Kong
London
Milan
Paris
Singapore
Tokyo

Dietmar Richter (Ed.)

Cell Polarity and Subcellular RNA Localization

With 44 Figures

 Springer

Professor Dr. DIETMAR RICHTER
Institut für Zellbiochemie
und klinische Neurobiologie
Universitäts-Krankenhaus Eppendorf
Martinistraße 52
20246 Hamburg
Germany

ISSN 0080-1844
ISBN 3-540-41142-9 Springer-Verlag Berlin Heidelberg New York

Library of Congress Cataloging-in-Publication Data

Cell polarity and subcellular RNA localization / Dietmar Richter (ed.).
    p. cm. – (Results and problems in cell differentiation ; 34)
    Includes bibliographical references and index.
    ISBN 3540411429
    1. RNA – Physiological transport. 2. Carrier proteins. 3. Polarity (Biology) 4.
Cytoskeletal proteins. I. Richter, Dietmar, 1939– II. Series.
    QP623 .C455 2001
    572'.69 – dc21

Springer-Verlag Berlin Heidelberg New York
a member of BertelsmannSpringer Science+Business Media GmbH

© Springer-Verlag Berlin Heidelberg 2001
Printed in Germany

Cover design: Meta Design, Berlin
Typesetting: Best-set Typesetter Ltd., Hong Kong
SPIN 10732748   39/3130 – 5 4 3 2 1 0 – Printed on acid-free paper

Preface

Cell polarity is reflected by an asymmetric distribution of macromolecules. Intracellular transport and the specific localization of proteins have been implicated in the establishment and maintenance of cellular polarity and plasticity. Protein localization mechanisms have been well characterized, often accomplished by cellular sorting systems that recognize intrinsic targeting signals, as have the modes of action of various motor proteins that underlie such mechanisms. It has become clear over the past decade that the differential subcellular localization of mRNAs represents an additional mechanism to achieve accumulation of cognate proteins at discrete sites within a cell.

This has been particularly well documented in *Drosophila*, but also in *Xenopus* oocytes. More recently, subcellular RNA transport has also been seen in nerve cells, allowing them to respond appropriately to external stimuli. Transport of organelles and proteins to their ultimate destinations at pre- and post-synaptic sites has long been thought to constitute the sole determinants of nerve cell polarity and complexity. This view has been changed, however, by the discovery of certain RNA species in neurites of a variety of nerve cells. Yet, details of the mechanism for subcellular RNA localization, including proteins as RNA-transporting molecular motors have remained elusive.

The objectives of this book are to review a few selected examples of the RNA kinesis field and to make clear to the reader the current state of the art and what unresolved problems remain. The chapters deal with various cellular systems including those of *Xenopus* oocytes, the mammalian brain, and the invertebrate nervous system, and describe subcellular RNA localization and translation, RNA targeting sequences (*cis*-elements), RNA-transporting proteins (*trans*-factors), activity-dependent regulation via transcription factors, and the involvement of the cytoskeleton in the context of neuronal function and long-term synaptic plasticity. The *Drosophila* system has not been included, mainly because it has been covered extensively in other series. Chapters also include nucleocytoplasmic RNA transport and the nuclear protein factors involved – a protein family that has recently been found to participate in the transport of mRNA outside the nucleus. Members of this family apparently bind specifically to *cis*-elements of targeted mRNAs, thus directing these mRNAs to a discrete subcellular compartment.

The book was initiated in the spring of 1999, while the editor stayed at the Neurosciences Institute, San Diego, California. At this beautiful, almost monastic, place with impressive scholarly resources, scientific visitors enjoy many discussions and frequent interactions with members of the Institute. It was at the end of a small workshop on 'Regulation of protein synthesis in nerve cells' that, after a stimulating discourse on the role of dendritic mRNA and its translation in neurons, Gerald Edelman rightly stated that although many problems remained to be solved, it will certainly be a challenging and exciting field for years to come. It is hoped that some of this excitement is transmitted by this book to the reader.

Dietmar Richter

Contents

Localization of mRNAs at Synaptic Sites on Dendrites
Oswald Steward and Paul Worley

Extrasomatic Targeting of MAP2, Vasopressin and Oxytocin mRNAs in Mammalian Neurons

Stefan Kindler, Evita Mohr, Monika Rehbein, and Dietmar Richter

RNA Transport and Local Protein Synthesis in the Dendritic Compartment

Alejandra Gardiol, Claudia Racca, and Antoine Triller

Neuronal BC1 RNA: Intracellular Transport
and Activity-Dependent Modulation
Jürgen Brosius and Henri Tiedge

Nucleocytoplasmic mRNA Transport
Yingqun Huang and Gordon G. Carmichael

RNA Localization in *Xenopus* Oocytes
Kinneret Rand and Joel Yisraeli

Local Protein Synthesis in Invertebrate Axons:
From Dogma to Dilemma
J. van Minnen and N.I. Syed

Nucleocytoplasmic RNA Transport in Retroviral Replication
Harald Wodrich and Hans-Georg Kräusslich

Localization of mRNAs at Synaptic Sites on Dendrites

Oswald Steward[1] and Paul Worley[2]

1 Introduction

It is becoming increasingly clear that an important aspect of gene expression
in neurons involves the delivery of mRNAs to particular subcellular domains.
Specifically, although the majority of the mRNAs that are expressed by neurons
are found only in the neuronal cell body, a select population of mRNAs are
transported into dendrites, and certain neurons also transport mRNA into
axons. The nature and significance of this RNA targeting is now under intense
investigation, and it is clear that mRNA targeting plays a key role in several
aspects of neuronal function.

The present chapter focuses on one aspect of RNA localization in neurons
– the localization of protein synthetic machinery and particular mRNAs at
synaptic sites on dendrites. We will briefly summarize what is known about
the protein synthetic machinery that is localized at synapses, update the list of
mRNAs that have been shown to be present in dendrites, and consider some
of the key principles regarding dendritic mRNAs. We will focus especially on
recent evidence regarding the mechanisms underlying mRNA sorting, transport, and selective localization at synapses as revealed by studies of the immediate early gene *Arc*. Finally, we will consider new information about the role
that local translation plays in synaptic function.

2 The Machinery for Translation in Dendrites

The story regarding the targeting of mRNAs to dendrites had its roots in the
discovery of synapse-associated polyribosome complexes (SPRCs). SPRCs are
polyribosomes and associated membranous cisterns that are selectively localized beneath postsynaptic sites on the dendrites of CNS neurons (Steward and
Levy 1982; Steward 1983; Steward and Fass 1983). Although it had long been
known that polyribosomes were present in dendrites, these studies in the early
1980s were the first to note and document the highly selective localization

[1] Reeve-Irvine Research Center, Departments of Anatomy/Neurobiology and Neurobiology and
Behavior, College of Medicine, University of California at Irvine, Irvine, CA 92697, USA
[2] Department of Neuroscience, Johns Hopkins University School of Medicine, Baltimore, MD
21205, USA

Results and Problems in Cell Differentiation, Vol. 34
D. Richter (Ed.): Cell Polarity and Subcellular RNA Localization
© Springer-Verlag Berlin Heidelberg 2001

beneath synapses. These studies also laid out two key working hypotheses: (1) that the machinery might synthesize key molecular constituents of the synapse and (2) that translation might be regulated by activity at the individual post-synaptic site. Studies over the last 18 years have confirmed and extended these hypotheses, and a coherent story regarding local protein synthesis at synaptic sites is now beginning to emerge. In considering this story, it is useful to begin by summarizing some of the key features of SPRCs.

2.1 SPRCs Are Precisely Localized in the Postsynaptic Cytoplasm

A key feature of SPRCs is the selectivity of their localization. Quantitative electron microscope analyses have revealed that the vast majority of the polyribosomes that are present in dendrites are *precisely* positioned beneath postsynaptic sites, and are absent from other parts of the dendrite (Steward and Levy 1982). SPRCs are most often localized at the base of the spine in the small mound-like structures from which the neck of the spine emerges. Thus, SPRCS are located within or near the portal between the spine neck and the shaft of the dendrite – the route through which current must flow when spine synapses are activated. In this location, SPRCs are ideally situated to be influenced by electrical and/or chemical signals from the synapse as well as by events within the dendrite proper. An important implication of this selective localization is that there must be some mechanism that causes ribosomes, mRNA, and other components of the translational machinery to dock selectively in the postsynaptic cytoplasm. The mechanisms underlying this highly selective localization remain to be established.

Although most dendritic polyribosomes are localized beneath synapses, a few clusters of ribosomes are localized within the core of the dendritic shaft. It is not yet known whether these represent a different population than the synapse-associated polyribosomes. One possibility is that the polyribosomes in the dendritic core are associated with mRNAs that encode proteins that are not destined for synaptic sites, but which play some other role in dendritic function. This speculation is of particular interest given the functional diversity of the mRNAs that have been identified in dendrites (see below). Alternatively, the clusters of polyribosomes in the core of the dendrite may represent packets of mRNAs and ribosomes that are in transit from the cell body.

2.2 SPRCs Are Present at Spine Synapses on Different Neuron Types

Quantitative analyses of polyribosomes have been carried out on dentate granule cells, hippocampal pyramidal cells, cortical neurons, and cerebellar Purkinje cells. These analyses reveal that SPRCs are present in a roughly similar configuration in all of the spine-bearing neurons that have been evaluated.

Estimates of the incidence of polyribosomes at spine synapses vary depending on the quantitative methods used. In evaluations of single sections, about 11–15% of the identified spines have underlying polyribosomes (Steward and

Levy 1982). However, this is clearly an underestimate because not all of the area under a spine is contained within a single section. For example, serial section reconstructions of dendrites in the dentate gyrus reveal that the actual incidence of polyribosomes in spines on mid proximo-distal dendrites is about 25% (Steward and Levy 1982).

The estimates of incidence also depend on the counting criteria. Studies that have used serial section reconstruction techniques to evaluate the distribution of individual ribosomes (not polyribosomes) yield higher estimates of incidence (Spacek and Hartmann 1983). For example, in pyramidal neurons in the cerebral cortex, 82% of the reconstructed spines had ribosomes in the head, 42% had ribosomes in the neck, and 62% had ribosomes at the base. In cerebellar Purkinje cells, 13% of the spines had ribosomes in the head and 22% had ribosomes at the base. It is likely that an important reason for the higher incidence values in this study is that singlet ribosomes were counted along with polyribosomes. In any case, it is clear that polyribosomes are a ubiquitous component of the postsynaptic cytoplasm in a variety of neuron types.

2.3 SPRCs Are Often Associated with Membranous Organelles in an RER-Like Configuration

Serial section reconstructions of dendrites of dentate granule cells and hippocampal pyramidal cells have revealed that about 50% of the polyribosomes are associated with tubular cisterns (Steward and Reeves 1988). A common configuration is one in which the ribosomes surround a blind end of a cistern. These configurations suggested that the SPRC/cisternal complex may be a form of RER that could allow the synthesis of integral membrane proteins or soluble proteins destined for release. In support of this hypothesis, recent studies have provided evidence for RER-associated proteins and also proteins of the Golgi apparatus in dendrites, some of which appear to be localized in these sub-synaptic cisterns (see below).

Interestingly, the cisterns with which SPRCs are associated are sometimes connected with a spine apparatus (Steward and Reeves 1988). The significance of these connections is not known. One interesting hypothesis is that the spine apparatus may involved in some aspect of post-translational processing of proteins that are synthesized at SPRCs. So far, however, there is no definitive evidence in support of this hypothesis.

2.4 SPRCs Are Also Present at Non-Spine Synapses

There have been no detailed quantitative evaluations of polyribosome distribution in the dendrites of non-spiny neurons. Nevertheless, it is clear that the same basic relationships exist as in spiny dendrites. For example, polyribosomes are often present beneath shaft synapses in association with sub-membranous cisterns, and are found beneath both asymmetric (presumed excitatory) and symmetric (presumed inhibitory) synapses.

Polyribosomes are also localized beneath synapses on axon initial segments (Steward and Ribak 1986). This localization is noteworthy for two reasons. First, it extends the generality of SPRCs to yet another type of postsynaptic location. Second, most (perhaps all) synapses on axon initial segments use GABA as their neurotransmitter, and are thus inhibitory.

It is noteworthy that axon initial segments also contain organelles that appear identical to spine apparatuses. When in the axon, such organelles are termed cisternal organelles. Because of the similarity in appearance between spine apparatuses and cisternal organelles beneath synapses on axon initial segments, we suggest a new term that would apply to both – *subsynaptic cisternal organelles*. Based on the localization of subsynaptic cisternal organelles beneath both excitatory and inhibitory synapses on axon initial segments, it may be worthwhile to reconsider the possible functions of these enigmatic organelles. This is especially true because previous hypotheses have focused on functions that would be especially important at excitatory synapses and perhaps of minimal importance at inhibitory synapses (for example Ca^{2+} sequestration).

2.5 SPRCs Are Especially Prominent at Developing Synapses

If protein synthetic machinery is localized at synapses in order to synthesize some of the components of the synaptic junction, one would expect SPRCs to be especially prominent at synapses during periods of synapse growth. This is the case. Polyribosomes are very abundant in the dendrites of developing neurons, and again appear to be preferentially localized beneath postsynaptic sites, although the degree of selectivity has not been evaluated quantitatively (Steward and Falk 1986).

3 Types of Proteins That Are Synthesized at SPRCs

The discovery of SPRCs raised the question of what proteins were synthesized in the postsynaptic cytoplasm. The approaches that have been used to address this question include:

1. Biochemical studies of proteins synthesized by subcellular fractions enriched in pinched-off dendrites;
2. In situ hybridization analyses of the subcellular distribution of mRNAs in neurons; and
3. Molecular biological analyses of the complement of mRNAs in isolated dendrites from immature neurons grown in culture.

3.1 Biochemical Studies of Proteins Synthesized by Subcellular Fractions Enriched in Pinched-Off Dendrites

Biochemical approaches take advantage of subcellular fractionation techniques that allow the isolation of synaptosomes with attached fragments of dendrites that retain their cytoplasmic constituents, including polyribosomes and associated mRNAs. These fractions were initially called heavy synaptosomes (Verity et al. 1980), but we prefer the term synaptodendrosomes to emphasize the fact that the fractions contain pinched-off dendrites (Rao and Steward 1991a). Others have used similar fractions prepared by filtration, which are termed synaptoneurosomes (Weiler and Greenough 1991, 1993; Weiler et al. 1997).

The major limitation in using synaptodendrosome or synaptoneurosome fractions to study dendritic protein synthesis is that the fractions are contaminated with fragments of neuronal and glial cell bodies. For example, high levels of the mRNA encoding glial fibrillary acidic protein are present (Chicurel et al. 1990; Rao and Steward 1991b), and it is likely that there are also fragments of neuronal cell bodies that contain mRNAs that are normally not present in dendrites.

One way to circumvent the problem of contamination is to focus on proteins that are synthesized in these fractions and then assembled into synaptic structures. For example, when synaptosomes are incubated with $[^{35}S]$-methionine, a number of protein species become labeled. Subcellular fractionation and detergent extraction techniques can then be used to prepare synaptic plasma membranes and fractions enriched in synaptic junctional complexes (the postsynaptic membrane specialization and associated membrane). Polyacrylamide gel electrophoresis combined with fluorography can then be used to reveal the proteins that were synthesized within the synaptodendrosomes and that had become associated with synaptic plasma membrane and synaptic junctional complex. This strategy has revealed characteristics of the labeled bands (Rao and Steward 1991a; Leski and Steward 1996), but so far, the approach has not provided definitive identification of the proteins. This combined strategy also has the limitation that it is only useful for proteins that are assembled into the synaptic membrane or synaptic junctional complex. Thus, proteins that are not assembled into the synapse are not detected.

Studies using synaptoneurosome fractions without the secondary purification step of subcellular fractionation have provided evidence for the dendritic synthesis of one novel protein that had not previously been identified – fragile X mental retardation protein (FMRP; Weiler et al. 1997). FMRP is encoded by a gene called *fmr1*, which is affected in human fragile X syndrome. Treatment of synaptoneurosomes with agonists for metabotropic glutamate receptors cause a rapid increase in the amount of FMRP in the synaptoneurosome fractions by Western blot analysis. These data suggested that FMRP was being synthesized within the fractions, and that the synthesis was enhanced by mGluR activation. This evidence has led to the interesting idea that the neuronal dysfunction that is part of fragile X syndrome may result from a disruption of local synthesis of protein at synapses (Comery et al. 1997; Weiler et al. 1997).

There are some inconsistencies in the story regarding FMRP, however. In the first place, FMRP mRNA does not appear to be present in dendrites at detectable levels (Hinds et al. 1993; Valentine et al. 2000). One can conceive of reasons why a particular mRNA might not be detected by in situ hybridization. For example, the mRNA could be present, but at levels that are below the threshold for detection by standard in situ hybridization techniques. This possibility is difficult to reconcile with the biochemical data, however, which indicate an almost twofold increase in the amount of FMRP detectable by Western blots within 5 min after treatment with mGluR agonists (Weiler et al. 1997). It is difficult to imagine how such a large change in protein concentration could be achieved unless the levels of the mRNA were very high.

It has also been suggested that FMRP plays a role in the regulation of translation of mRNAs at synapses (Feng et al. 1997). This hypothesis is based on two facts: (1) FMRP is an RNA binding protein and (2) immunocytochemical studies reveal that the protein is localized at polyribosome clusters. Of course, this localization would also be consistent with the hypothesis that FMRP is synthesized within dendrites. It remains to be seen how the story regarding local synthesis of FMRP will unfold.

The only other protein that has been definitively identified as being synthesized in synaptoneurosomes is the alpha subunit of calcium/calmodulin-dependent protein kinase II (CAMKII; Sheetz et al. 2000). These studies involved metabolic labeling with $[^{35}S]$-methionine followed by 2-D gel electrophoresis and fluorography or by immunoprecipitation of metabolically-labeled protein using an antibody for CAMKII. Of course, it was already known that CAMKII mRNA was present in dendrites before this study was undertaken. Indeed, the purpose of this study was to evaluate whether CAMKII synthesis was regulated by neurotransmitters (see below). Still, the study documents that certain mRNAs known to be present in dendrites are in fact translated in synaptoneurosomes.

3.2 mRNAs in the Dendrites of Neurons in Vivo

Important clues about the identity of the proteins that may be synthesized at SPRCs has come from in situ hybridization analyses that document the presence of particular mRNAs in dendrites. In most studies, dendritic localization has been inferred by the pattern of labeling in brain regions where neurons are collected in discrete layers, and where there are distinct neuropil layers that contain dendrites and axons but few neuronal cell bodies (cortical regions including the hippocampus and the cerebellar cortex). Definitive evidence that the mRNA is in fact present in dendrites (and not in glial cells) can be obtained using non-isotopic in situ hybridization techniques. Dendritic localization can also be confirmed by studies of neurons in culture, although one must consider the possibility that neurons in culture may express an unusual complement of mRNAs or sort mRNAs in different ways than neurons in vivo.

One important caveat is that the presence of a particular mRNA in dendrites does not establish that the mRNA is translated at SPRCs. There are polyribosomes in dendrites which are not localized beneath synapses, and these polyribosomes could be associated with a different set of mRNAs than are translated in the postsynaptic cytoplasm (Steward and Reeves 1988). Nevertheless, identification of mRNAs that are present in dendrites provides candidates for synapse-associated mRNAs that can be further evaluated in other ways.

Table 1 lists the mRNAs for which the evidence for dendritic localization in vivo is strong. All of the RNAs listed extend for several hundred micrometers from the cell body. Certain other mRNAs that are localized primarily in cell bodies may extend slightly into proximal dendrites. For example, it has been reported that the mRNA for two protein kinase C substrates (F1/GAP43 and RC3) extend somewhat further into the proximal dendrites of forebrain than other cell body mRNAs (Laudry et al. 1993). The differences in the distribution of mRNAs encoding F1/GAP43 and RC3 vs other cell body mRNAs are slight, and indeed were not evident in studies using non-isotopic in situ hybridization techniques that produced heavy labeling over cell bodies (Paradies and Steward 1997).

Taken together, the information on mRNAs in dendrites allows several generalizations:

1. *A number of different mRNAs that encode seemingly unrelated proteins are present in dendrites.* The proteins encoded by mRNAs that are present in dendrites include a variety of different classes of protein (Table 1), including cytoplasmic, cytoskeletal, integral membrane, and membrane-associated proteins. The proteins also have very different functions. Thus, it is unlikely that there will be a single purpose for mRNA localization in dendrites (for additional discussion of this point, see Steward (1997).
2. *All of the dendritic mRNAs that have been identified so far are expressed differentially by different types of neurons.* This is especially evident when considering the mRNAs that are present in the dendrites of forebrain neurons vs cerebellar Purkinje cells. For example, the mRNAs for MAP2, CAMKII, den, and Arc are found in forebrain neurons, but are not expressed at high levels by Purkinje cells. Purkinje cells, on the other hand, express a different complement of mRNAs including the IP3 receptor and other Ca^{2+}-interacting proteins (for example, L7). These mRNAs are expressed by Purkinje cells and a few other neuron types. The fact that a different mixture of mRNAs is present in the dendrites of different cell types suggests that dendritic protein synthesis may have different purposes in different cell types.
3. *Although a number of mRNAs are present in the dendrites of forebrain neurons, the patterns of expression and subcellular distributions of the mRNAs are different.* The mRNAs for CAMKII, dendrin, and Arc (when induced) are localized throughout the dendrites. In contrast, the mRNA for MAP2 is found at high levels in the proximal one third to one half of the total dendritic length, but is not detectable in distal dendrites of most neurons. The mRNAs that are present in the dendrites of Purkinje cells also exhibit different localization patterns. For example, the mRNA for the IP3 receptor is present throughout dendrites but is concentrated in the proximal one third of the total dendritic length. The mRNA for L7 appears to be

Table 1. mRNAs that have been shown to be localized within dendrites of neurons in vivo by in situ hybridization. Not shown are mRNAs that are localized only in the most proximal segments

mRNA	Cell type	Localization in dendrites	Class of protein	Protein function	Reference
MAP2	Cortex, hippocampus dentate gyrus	Proximal one third to one half	Cytoskeletal	Microtubule-associated	Garner et al. (1988)
CAMII kinase alpha subunit	Cortex, hippocampus dentate gyrus	Throughout	Membrane-associated postsynaptic density	Multifunctional kinase Ca^{2+} signaling	Burgin et al. (1990)
Arc/Arg 3.1	Cortex, hippocampus, dentate gyrus depending on inducing stimulus	Throughout (when induced)	Cytoskeleton-associated	Actin-binding ?? synaptic junctional protein	Link et al. (1995); Lyford et al. (1995)
Dendrin	Hippocampus, dentate gyrus, cerebral cortex	Throughout	Putative membrane	Unknown	Herb et al. (1997)
G-Protein gamma subunit	Cortex, hippocampus, dentate gyrus, striatum	Throughout	Membrane-associated	Metabotrophic receptor signaling	Watson et al. (1994)
Calmodulin	Cortex, hippocampus, Purkinje cells	Proximal–middle (during synaptogenesis)	Cytoplasm and membrane-associated	Ca^{2+} signaling in conjunction with CAMII kinase	Berry and Brown (1996)
NMDAR1	Dentate gyrus	Proximal–middle?	Integral membrane	Receptor	Benson (1997)
Glycine receptor alpha subunit	Motoneurons	Proximal	Integral membrane	Receptor	Racca et al. (1997)
Vasopressin	Hypothalamo-hypophyseal	Proximal–middle	Soluble	Neuropeptide	Prakash et al. (1997)
Neurofilament protein 68	Vestibular neurons	Proximal–middle	Cytoskeletal	Neurofilament	Paradies and Steward (1997)
InsP3 receptor	Purkinje cells	Throughout (concentrated proximally)	Integral membrane (endoplasmic reticulum)	Ca^{2+} signaling	Furuichi et al. (1993)
L7	Purkinje cells	Throughout	Cytoplasmic?	Homology to c-sis PDGF oncogene signaling?	Bian et al. (1996)
PEP19	Purkinje cells	Proximal one third	Cytoplasmic	Ca^{2+}-binding	Furuichi et al. (1993)

Table 2. Differential distribution of mRNAs in dendrites: Relationship to synapse type on neurons in the hippocampus

Cell Type	mRNA			
	CAMKII	MAP2	Den	Arc
Dentate granule cell				
Proximal (Comm/Assoc)[a]	+++	+++	+++	+++
Middle (MPP)[b]	+++	++	+++	+++
Distal (LPP)[c]	+++		+++	+++
CA1 pyramidal cell				
Proximal (Comm/Assoc)	+++	+++	++	++
Middle (Comm/Assoc)	+++	+	++	++
Distal (temporo-ammonic)	+++		++	++
CA3 pyramidal cell				
Proximal (mossy fiber)	+	+		++
Middle (Comm/Assoc)	+++	++		++
Distal (MPP/LPP)	+++			++

[a] Comm/Assoc = Commissural/associational
[b] MPP = medial perforant path
[c] LPP = Lateral perforant path

more uniformly distributed throughout the dendrites (Bian et al. 1996). Studies of the trafficking of the immediate early gene Arc also reveal that the complement of dendritic mRNAs can vary over time in an activity-dependent fashion (see below). These findings indicate that the capability exists for a different mixture of proteins to be synthesized locally at different times in different neuron types and different dendritic domains, providing a considerable complexity in the mechanisms underlying gene expression at dendritic synapses. Also, the variety of subcellular distributions implies that there must be multiple signals mediating mRNA localization within dendrites.

One interesting implication of these data is that it is now possible to begin to construct data bases of the sort illustrated in Table 2, which summarizes the complement of mRNAs that are present in different dendritic domains of particular neuron types. For some of these neurons (hippocampal neurons and neurons of the dentate gyrus), different types of synapses terminate selectively at particular locations on the dendrite. Moreover, the different types of synapses have different physiological properties. Thus, it may eventually be possible to determine whether the complement of mRNAs that are present in a particular dendritic segment determines the physiological properties of the synapses that terminate on those segments.

4. *The presence of mRNAs encoding different classes of proteins implies the existence of different types of translational machinery in dendrites.* MAP2, CAMKII, and Arc are non-membrane proteins and thus would presumably

be synthesized by 'free polysomes'. In contrast, the InsP3 receptor is an integral membrane protein that presumably must be synthesized by membrane-bound ribosomes (rough endoplasmic reticulum, RER). As noted above, electron microscopic studies have revealed subsynaptic polyribosomes closely associated with membranous cisterns that may represent a form of RER (Steward and Reeves 1988). However, the InsP3 receptor is also a glycoprotein, and so there is the question of how the newly synthesized protein is glycosylated.

There are several other mRNAs that are detectable by in situ hybridization in the dendrites of young neurons that are not evident in the dendrites of mature neurons in vivo. For example, the mRNA for calmodulin can be detected in dendritic laminae in developing animals, but not in mature ones (Berry and Brown 1996). This is consistent with the idea that local dendritic protein synthesis is especially important during periods of synaptogenesis (Palacios-Pru et al. 1981, 1988; Steward and Falk 1986).

Several other mRNAs have been shown to be present in the dendrites of young neurons developing in vitro. For example, the mRNAs for BDNF and trkB receptors extend into the proximal 30% of the total dendritic length of hippocampal neurons in culture (Tongiorgi et al. 1997). Potassium-induced depolarization increases the extent of dendritic labeling, so that the mRNAs extend to an average of 68% of the total dendritic length. Despite the easily-detectable dendritic labeling in neurons in vivo, the mRNAs for BDNF and trkB receptors appear to be largely restricted to the region of the cell body in young neurons in vivo (Dugich et al. 1992). It remains to be seen whether a dendritic localization could be induced in neurons in vivo by manipulating neuronal activity.

Another mRNA that has been demonstrated in the dendrites of hippocampal neurons in vitro encodes the fatty-acylated membrane-bound protein ligatin. Fluorescent in situ hybridization analyses indicate that the mRNA for ligatin extends for over 100 μm into the dendrites of hippocampal neurons in culture, producing dendritic labeling that is nearly as extensive as that produced by probes for CAMKII (Severt et al. 2000). Nevertheless, previous studies of ligatin mRNA distribution in neurons in vivo reveal that the mRNA is largely restricted to the cell body region (Perlin et al. 1993). It is possible that the radio-isotopic in situ hybridization techniques used in the earlier study in vivo were not sufficiently sensitive to detect mRNA in dendritic laminae. In any case, the issue of whether ligatin mRNA is present in dendrites in vivo remains to be resolved.

3.3 Molecular Biological Analyses of the Complement of mRNAs in Isolated Dendrites from Neurons in Vitro

One approach to identifying dendritic mRNAs has been to use patch pipettes to aspirate the cytoplasmic contents of individual dendrites of neurons grown in culture and then use RNA amplification techniques to clone the mRNAs (Miyashiro et al. 1994). This study provided intriguing evidence that there may be a substantial number of mRNAs in dendrites, many of which remain to be characterized. However, there are certain inconsistencies between these and other findings. For example, the mRNAs for GluR1 receptors were detected, although the mRNAs for these receptors have not been detected in dendrites of neurons in vivo or in vitro. The reason for the disparity of results is not clear.

There are several possible explanations for the inconsistencies. One possibility is that Miyashiro et al. analyzed cytoplasm from dendrites of very young neurons developing in culture. These might contain a different complement of mRNAs than the dendrites of mature neurons. Another possibility is that the amplification techniques detect mRNAs which are present in such low abundance that they are not easily detected using routine in situ hybridization techniques. It is also conceivable that some of the mRNAs in dendrites are in a form that somehow interferes with hybridization by complementary probes. In any case, it seems prudent to confirm the dendritic localization of any mRNAs detected in this way by in situ hybridization before drawing firm conclusions about dendritic localization.

3.4 The Search for Novel Dendritic mRNAs

It is almost certain that there are more dendritic mRNAs yet to be found. For example, biochemical studies of proteins synthesized within synaptodendrosomes suggest that several as yet unidentified constituents of synaptic junctions are locally synthesized within dendrites (Rao and Steward 1991a; Steward et al. 1991). The most prominent of these are not in a molecular weight range that would be consistent with their being the translation products of known dendritic mRNAs.

Systematic searches for new members of the family of dendritic mRNAs must deal with the problem of how to obtain sufficient quantities of mRNA from dendrites that are not contaminated by mRNA from neuronal cell bodies or supporting cells. So far, systematic searches have not yet identified new members of the family of dendritic mRNAs whose presence in dendrites in vivo was confirmed by in situ hybridization analyses. Given this past history, it seems likely that new members of the family will be found by chance through in situ hybridization analyses of the transcripts of newly cloned genes.

4 Activity-Dependent Regulation of mRNA Trafficking in Dendrites

There is now clear evidence that synaptic activation triggers the delivery of new mRNA transcripts to the synapse. This evidence has come from studies of the immediate early gene *Arc*. *Arc* was discovered in screens for novel immediate early genes (IEGs), defined as genes that are induced by neuronal activity in a protein synthesis-independent fashion (Link et al. 1995; Lyford et al.

1995). In both studies, the inducing stimulus was a single electro-convulsive seizure (ECS), and the protein synthesis-independence was ensured by treating animals with cycloheximide to block protein synthesis (and hence the synthesis of secondary response genes). *Arc* was one of a number of novel IEGs that were identified using this paradigm, but was unique in that all other IEG mRNAs remained in the cell body, whereas *Arc* mRNA rapidly migrated into dendrites.

The expression dynamics of *Arc* provide unique advantages for studies of mRNA trafficking because it is expressed as an immediate early gene (IEG). *Arc* mRNA is expressed at very low levels constitutively. In fact, in resting control animals, only a few neurons express *Arc* at detectable levels (see for example Fig. 1A). Nevertheless, *Arc* is strongly induced by neuronal activity. Hence, the synthesis, intracellular transport, localization, and life history of *Arc* mRNA can be studied in a way that is not possible with mRNAs that are expressed constitutively. For example, Fig. 1 illustrates the distribution of *Arc* mRNA in the dentate gyrus of control animals (A) and 2h after a single ECS (B). The remarkable feature of *Arc* is that the mRNA is delivered throughout dendrites. Indeed, newly-synthesized *Arc* mRNA reaches the most distal tips of the dendrites of dentate granule cells within 1 h after the inducing stimulus (Wallace et al. 1998). It is about 300 μm from the granule cell body to the distal tips of the dendrites; thus, *Arc* mRNA migrates into dendrites at a rate of at

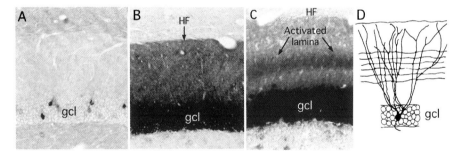

Fig. 1A–D. Selective dendritic targeting of *Arc* mRNA. The photomicrographs illustrate the distribution of *Arc* mRNA in the dentate gyrus as revealed by non-isotopic in situ hybridization. **A** Distribution of *Arc* mRNA in the dentate gyrus in an animal that had been left undisturbed in its home cage for the period immediately preceding euthanasia. **B** Induction of *Arc* and dendritic delivery of *Arc* mRNA 2h after a single electroconvulsive seizure (ECS). **C** Distribution of *Arc* mRNA in the dentate gyrus after high frequency stimulation of the perforant path. **B, C** Illustrate the results of an experiment in which a single ECS was delivered to induce *Arc* expression bilaterally. Then, the rat was anesthetized, and stimulation and recording electrodes were positioned so as to activate the medial perforant path on the right side. The band of labeling occurs in the dendritic zone in which the perforant path terminates. **D** The drawing illustrates the distribution of the dendrites of the dentate granule cells in the molecular layer, and the pattern of termination of the medial perforant path projections from the entorhinal cortex;*gcl* granule cell layer. Panels **A, B,** and **C** are from Steward and Worley (submitted)

least 300 μm/h. These observations provided the first estimates of the rate of migration of a particular dendritic mRNA (Wallace et al. 1998).

Newly Synthesized Arc mRNA Is Selectively Targeted to Activated Synapses. An even more remarkable feature of Arc is that the newly synthesized mRNA is selectively targeted to active synaptic sites (Steward et al. 1998). It was found that patterned synaptic activation both induces *Arc* and causes the newly-synthesized mRNA to localize selectively in activated dendritic domains. This was demonstrated in studies in which *Arc* was induced by stimulating the entorhinal cortical projections to the dentate gyrus in anesthetized rats. The projection from the entorhinal cortex to the dentate gyrus (the perforant path) terminates in a topographically-organized fashion along the dendrites of dentate granule cells. Projections from the medial entorhinal cortex terminate in the middle molecular layer of the dentate gyrus whereas projections from the lateral entorhinal cortex terminate in the outer molecular layer. Hence, by positioning a stimulating electrode in the medial EC, it is possible to selectively activate a band of synapses that terminate on mid proximo-distal dendrites.

When the medial perforant path was activated using a stimulation paradigm that is typically used to induce LTP (400 Hz trains, eight pulses per train, delivered at a rate of one every 10 s), *Arc* was strongly induced. The newly-synthesized mRNA migrated into dendrites, and accumulated selectively in the middle molecular layer in exactly the location of the band of synapses that had been activated (Fig. 1C).

Similarly, activation of other afferent systems that terminate at different locations in the molecular layer caused the newly-synthesized mRNA to localize selectively in other dendritic laminae. For example, simulation of the lateral entorhinal cortex produced a band of labeling for *Arc* mRNA in the outer molecular layer; stimulation of the commissural projection produced a band of labeling in the inner molecular layer (Steward et al. 1998). These experiments thus revealed that synaptic activation generated some signal that caused Arc mRNA to localize near the active synapses. The nature of this docking signal remains to be defined.

Localization of Arc mRNA in Activated Dendritic Laminae Is Associated with a Local Accumulation of Arc Protein. Immunostaining of tissue sections from stimulated animals using an Arc-specific antibody revealed a band of newly-synthesized protein in the same dendritic laminae in which Arc mRNA was concentrated (Steward et al. 1998). The fact that synaptic activation leads to the selective targeting of both recently synthesized mRNA and protein suggests that the targeting of the mRNA underlies a local synthesis of the protein.

One additional important point revealed by immunocytochemistry is that newly-synthesized Arc protein is also present in the neuronal cell body, and all along the proximal dendrites of the dentate granule cells. Thus, it is not the case that *Arc* mRNA is translationally repressed until it reaches the activated dendritic domain.

4.1 Lessons from the Study of *Arc*: A Cellular Mechanism for Protein Synthesis-Dependent Synaptic Modification

The selective targeting of the newly synthesized mRNA to synapses that had been activated reveals the existence of a previously unknown mechanism that is ideally suited to mediate protein synthesis-dependent, synapse-specific, long-term plasticity. The fact that newly synthesized Arc protein accumulates selectively in the dendritic laminae in which *Arc* mRNA is localized also suggests that targeting of the mRNA underlies a local synthesis of the protein. The key points are:

1. The expression of the *Arc* gene is triggered by the patterns of synaptic activity that lead to long-term synaptic modification,
2. *Arc* mRNA is selectively targeted to synapses that have experienced a particular pattern of activity,
3. Arc protein accumulates in postsynaptic junctions, and
4. The activity-dependent induction of *Arc* occurs during a time window that extends for a few hours after the inducing stimulus.

There are a number of pieces of the puzzle that are still missing, however. First, it remains to be established whether Arc is directly involved in activity-induced synaptic modification. Additional clues about the actual role of the protein will likely come from studies of the protein itself, and its interactions with other functional molecules of the synaptic junctional region. Whatever its role in synaptic plasticity, the way that Arc is handled by neurons reveals the existence of RNA trafficking mechanisms that could be used for sorting other mRNAs that do play a key role in bringing about activity-dependent modifications.

The fascinating properties of Arc should not make us lose sight of the fact that other mRNAs are present in dendrites constitutively, including the mRNAs for molecules that have already been strongly implicated in activity-dependent synaptic modification (the mRNA for the alpha subunit of CAMII kinase for example). These mRNAs that are present constitutively provide an opportunity for local regulation of the synthesis of key signaling molecules via translational regulation (see above). Hence, gene expression at individual synapses may be regulated in a complex fashion, first through the regulation of the mRNAs available for translation (i.e. *Arc*) and then by regulation of the translation of the mix of mRNAs that are in place, including those present constitutively. How this is coordinated remains to be established.

4.2 Post-Translational Processing Within Dendrites

The presence of mRNAs encoding integral membrane proteins raises the question of whether dendrites contain the machinery for post-translational processing of recently synthesized proteins (specifically, components of the RER and Golgi apparatus). This question has been evaluated by assessing the distribution of protein markers and enzyme activities characteristic of the RER and Golgi apparatus (especially glycosyl-transferase activities).

Initial studies evaluated the distribution of glycosyl-transferase activities by pulse-labeling neurons with the sugar precursors which are the substrates of various glycosyl-transferases that are present in the RER and Golgi apparatus (GA). For example, mannose is added to nascent glycoproteins in the RER. Thus, when neurons are pulse-labeled with ^{3}H-mannose under conditions in which transport of recently synthesized proteins is blocked, the sites of mannose incorporation can be revealed autoradiographically. The same strategy can be applied to higher order glycosyltransferases that are characteristic of the GA, for example fucosyltransferase and galactosyltransferase. Studies of this type provided the initial evidence for the presence of both mannosyl-transferase as well as higher-order (Golgi-like) glycosyltransferase activity in the dendrites of hippocampal neurons grown in culture (Torre and Steward 1996).

Additional evidence regarding the localization of the RER and GA in dendrites has come from immunocytochemical studies of the subcellular distribution of proteins that are considered markers of the two endomembrane systems (Torre and Steward 1996). Immunocytochemical studies of the distribution of ribophorin I (an RER marker) reveal staining that extends well into dendrites. In general, immunostaining for Golgi markers extended only into proximal dendrites, however. The immunocytochemical data were generally consistent with the autoradiographic evidence regarding the intracellular distribution of glycosyltransferase activity characteristic of RER and Golgi.

These data raise the question of what membranous organelles are actually responsible for the activities characteristic of the RER and GA. Recent electron microscope immunocytochemical studies indicate that the membranous cisterns that are present near spine synapses stain for Sec6Ialpha protein complex, which is part of the machinery for translocation of proteins through the RER during their synthesis (Pierce et al. 2000). The cisterns exhibiting labeling have the same appearance as the cisterns that represent the membranous component of SPRCs. Moreover, immunostaining for the ribosomal protein S3 revealed labeling over the same membranous cisterns that were labeled for Sec6Ialpha.

The organelle responsible for higher-order glycosyltransferase (Golgi-like) activity remains to be identified. The studies of RER and GA distribution in dendrites imply that RER is present throughout dendrites whereas machinery for higher order glycosylation is present only in proximal dendrites, at least in hippocampal neurons in culture (Torre and Steward 1996). Thus, if any integral membrane proteins are synthesized in distal dendrites of forebrain neurons, their glycosylation may be incomplete. It will be of considerable interest to determine whether the story is different in Purkinje cells, where the mRNA for at least one integral membrane protein is present throughout dendrites.

5 Signals That Mediate mRNA Routing and Localization

The mRNAs that are delivered into dendrites represent a very small subset of the transcripts that a neuron expresses. Hence, there must be some mechanism for selectively targeting certain mRNAs for delivery into dendrites and/or retaining most mRNAs in the region of the cell body. Studies of other cell types in which mRNAs are differentially localized suggest that mRNA trafficking involves two distinct steps; transport from one site to another, followed by a docking process that mediates selective localization (Yisraeli et al. 1990; Steward and Singer 1997).

With the goal of developing a standard terminology, we propose the following terms to refer to the steps involved in mRNA trafficking and localization within neurons: (1) we propose the term routing to refer to the processes through which particular mRNAs are retained in the cell body or exported into dendrites or axons. The signals that direct the routing (which in other cell types are often located in the mRNA sequence) would then be termed routing signals. This signal has also been called an RNA zip code (Steward and Singer 1997). (2) Localization refers to the selective accumulation of mRNA at a particular intracellular site. Signals within the mRNA sequence or the encoded protein that mediate localization would then be termed localization signals, and the addresses that mark an intracellular site and cause a particular mRNA to localize selectively would be termed address markers.

In principle, routing and localization could be mediated by signals in the mRNA itself or signals in the encoded protein. In all of the examples known so far, however, the routing and localization appear to be based on signals in the mRNA itself.

Studies of the mechanisms of RNA localization in oocytes and embryos have provided the basic experimental paradigm for evaluating localization mechanisms. For example, one of the first localization signals to be characterized was the one responsible for localizing Vg1. The strategy was to construct chimeric mRNAs that include various segments of the localized mRNA, together with a reporter sequence, introduce these chimeric transcripts into cells, and then define the subcellular localization of the fusion transcripts using in situ hybridization. It was found that fusion transcripts containing the 3′ UTR of the Vg1 mRNA were localized like native Vg1 mRNA, whereas fusion transcripts lacking this region were not localized (Yisraeli and Melton 1988). Subsequent studies narrowed the signal to a segment of the 3′ untranslated region that is 340 nucleotides long (Mowry and Melton 1992). Similarly, it was found that the differential localization of bcd mRNA also depends on a signal in the 3′ untranslated region of the mRNA. Chimeric mRNAs containing a 630-nucleotide-long portion of the 3′ untranslated region were properly localized whereas localization was eliminated when 100–150 nucleotides were removed from either end of this segment (MacDonald and Struhl 1988).

The same strategies have now been applied to other localized mRNAs; in all cases so far, the localization signals turn out to be mediated by *cis*-acting signals in the 3′ UTR. This generalization holds for all of the localized mRNAs in oocytes and embryos, as well as for the mRNAs that have been characterized in non-neuronal cells. For example, the signals that are necessary and sufficient for localizing the mRNA for ß-actin in fibroblasts reside in a portion of the 3′ UTR (Kislauskis et al. 1994). In fact, there are multiple sequences that are capable of conferring localization in the 3′ UTR of ß-actin mRNA. One sequence is about 54 bases in length, and a second 43-nucleotide sequence exists 3′ of the 54-nucleotide sequence. These sequences are termed the zip code elements. Both the 54- and the 43-nt sequences can confer some localization (although the two are not equivalent in their activity). In oligodendrocytes, myelin basic protein mRNA is targeted for delivery from the cell body into processes by a signal that has been termed the RNA transport signal (RTS; Ainger et al. 1993, 1997). Another sequence, termed the RNA localization sequence (RLS) mediates the movement of the mRNA into the membrane lamellae at the distal end of the oligodendrocyte process. The finding that particular nucleotide sequences may mediate routing and localization gives added incentive for studies that seek to identify sequence homologies between different mRNAs that are targeted to similar subcellular domains.

In the case of dendritic mRNAs, the information is still rather sparse. The mRNA for *Arc* appears to be routed to dendrites by a signal in the mRNA itself. This conclusion is based on the fact that the movement of *Arc* mRNA into dendrites is not impeded when protein synthesis is blocked (Wallace et al. 1998). If targeting depended on the encoded protein, inhibition of protein synthesis should prevent the movement of the mRNA. Targeting of *Arc* mRNA to active synapses also continues in the presence of protein synthesis inhibitors, indicating that synapse-specific localization is also mediated by a signal in the mRNA (Steward et al. 1998). This type of approach provides no information, however, about the nature of the routing/localization signals, or where the signals reside in the mRNA sequence.

In the case of the mRNA for CAMKII, it has been shown that the 3′UTR is sufficient for dendritic routing. For example, transgenic animals have been created that express a transgene in which the 3′UTR of CAMKII is linked with a reporter (lacZ). This transcript is delivered into dendrites, and comes to be localized in a pattern that resembles native CAMKII mRNA (Mayford et al. 1996). The nature, length, and location of the signal within the 3′ UTR (which is over 4 kb in length) remain to be established.

In the case of MAP2 mRNA, the evidence suggests that the routing signal is contained within a 640-nucleotide sequence in the 3′ UTR (Blichenberg et al. 1999). The evidence in support of this conclusion comes from elegant experiments in which fusion transcripts were constructed from different parts of MAP2 mRNA together with a reporter sequence (the mRNA for green fluorescent protein). These were then introduced into hippocampal and sympathetic neurons in culture by microinjection or transfection, and the distribution of the transcripts was evaluated using in situ hybridization. When fusion transcripts that contained the 640-nucleotide sequence were expressed, about 50–60% of the transfected neurons exhibited dendritic labeling. When fusion transcripts containing other parts of the 3′UTR were expressed, a smaller proportion of neurons exhibited dendritic labeling (ranging from a few percent to 20% depending on the transcript and the neuron type). Thus, the authors concluded that the 640-nucleotide sequence was both necessary and sufficient for MAP2 mRNA localization.

It is of course premature to conclude that all examples of mRNA localization will be mediated by signals in the 3′ UTR, especially given the heterogeneity of dendritic mRNAs.

5.1 Granules: Supramolecular Packets That May Be Involved in mRNA Transport

Non-isotopic in situ hybridization techniques have revealed that many localized mRNAs appear to be present in some sort of supra-molecular aggregate that appears as a granule at the light microscope level. The first evidence that mRNA may be present in granules came from studies of the distribution of fluorescently labeled myelin basic protein mRNA after injection into oligodendrocytes (Ainger et al. 1993). It was found that the injected RNAs rapidly aggregated into granules, and these granules were then translocated into the oligodendrocyte processes. Studies of the distribution of native MBP mRNA revealed that this also appeared to be localized in granules.

There are now many examples of mRNAs in granules of one size or another, including the localized mRNAs in neurons (Knowles et al. 1996; Paradies and Steward 1997). The existence of granules implies that RNAs may be assembled into supra-molecular structures. Likely components of such supra-molecular assemblies include the mRNAs themselves (present as multiple copies?), RNA binding proteins that mediate routing, components of the RNA transport machinery, RNA-binding proteins that mediate docking of the mRNAs at their final destination, and components of the translational machinery, including ribosomes. Identifying the RNA-binding proteins present in the complex that mediate routing and/or localization is now a high priority.

In this regard, one interesting RNA binding protein is *staufen*. *Staufen* mediates the asymmetric distribution of key maternal mRNAs in *Drosophila* oocytes including the mRNAs for *bicoid* (which is targeted to the anterior pole) and *oskar* (which is localized to the posterior pole). *Staufen* has been shown to form granules upon micro injection into *Drosophila* oocytes (Ferrandon et al. 1994). Hence, there has been considerable interest in whether *staufen* or a related protein plays a role in mRNA localization in neurons. Mammalian *staufen*-related genes have now been cloned (Kiebler et al. 1999), and immunostaining experiments reveal that staufen protein is present in cell bodies and dendrites. In addition, staufen-positive granules that contain RNA move along microtubules in the dendrites (Kohrmann et al. 1999). These observations are consistent with a role for this protein in dendritic RNA routing/localization, but definitive evidence regarding the functional role of *staufen* in neurons is so far lacking. Also, it is important to recall the fact that different dendritic mRNAs have a different localization pattern, and so it seems likely that multiple mRNA binding proteins may be involved in bringing about this differential localization.

5.2 mRNA Translocation Mechanisms

The mechanisms through which mRNAs are delivered from the site of synthesis in the nucleus into dendrites remain to be established, but the evidence that is available suggests some form of active transport process.

Initial studies evaluated the rate of translocation of newly-synthesized RNA following pulse-labeling with ^{3}H-uridine. Autoradiographic analyses revealed that the newly synthesized RNA moved into dendrites at a rate of around 12 µm/h (Davis et al. 1987, 1990). In relatively mature neurons (2 weeks in vitro), there was little or no detectable transport into axons, although axonal transport was seen in young neurons prior to the time that they exhibit distinctive axons and dendrites (Kleiman et al. 1994). These data were somewhat un-satisfying, however, because the autoradiographic techniques provide low-resolution, and document bulk movement, much of which may represent newly-synthesized ribosomal RNA. Also, movement cannot be directly visualized in these types of experiments. Instead, movement is inferred by the change in distribution of the newly-synthesized RNA when neurons were fixed at various times after pulse labeling.

More recent studies have evaluated the actual movement of RNA-containing granules in the dendrites of neurons in culture (Knowles et al. 1996; Knowles and Kosik 1997). In these experiments, endogenous RNA was labeled using the membrane permeable nucleic acid dye SYTO-14, and the movement of RNA was evaluated in living neurons using video microscopy. The dye-labeled discrete granules in neuronal processes that were also positive for poly(A+) mRNA (as revealed by in situ hybridization), the ribosomal 60S subunit, and elongation factor 1-alpha (as revealed by immunocytochemistry). When the living neurons were observed over time, most of the granules did not move, suggesting that RNA granules exist primarily in a stationary phase. Nevertheless, about 3% of the granules were observed to move in a vectorial fashion within the dendrites at an average rate of 0.1 µm/s, which translates to a rate of 360 µm/h. Both anterograde and retrograde movements were seen in cultures older than 7 days. It is interesting that this is approximately the same rate of translocation seen with newly synthesized *Arc* mRNA in the dendrites of dentate granule cells in vivo (see above). Although informative, these experiments are still limited because the identity of the RNA that is labeled with SYTO-14 is not known. Thus, the rate of movement of any particular mRNA remains to be defined.

5.3 Regulation of mRNA Translation at Synapses

The selective localization of SPRCs at synapses provides a potential mechanism for locally regulating the production of certain key proteins based on
signals generated at the individual synapse. This local synthesis could involve
mRNAs already in place, and/or mRNAs that are induced by synaptic activity
and delivered into dendrites. Until recently, there was no evidence of a linkage
between synaptic activation and local translation. However, recent studies
establish that a linkage does exist, and that there may be complex signal transduction events that control the translation.

5.3.1 Studies in Hippocampal Slices

Initial evidence that synaptic activity might regulate dendritic protein synthesis came from autoradiographic studies of incorporation of labeled amino
acids in hippocampal slices (Feig and Lipton 1993). Activation of the Schaffer
collateral system (which terminates on the apical dendrites of hippocampal
neurons), in conjunction with activation of muscarinic acetylcholine receptors
by carbachol, caused increased labeling for newly-synthesized protein in
the activated dendritic laminae (stratum radiatum). The rapid appearance of
the increased labeling suggested local synthesis within dendrites; however, the
possibility of rapid transport of the newly synthesized proteins from the cell
body in these intact neurons cannot be excluded in these experiments.

More recent studies have provided evidence that synaptic activation causes
a local synthesis of proteins whose mRNAs are present constitutively in
dendrites (CAMKII and MAP2). Ouyang et al. (1998) evaluated changes in
immunostaining for CAMKII in hippocampal slices in which the Schaffer collateral system had been stimulated at frequencies that induce LTP. These
studies revealed rapid increases in immunostaining for CAMKII in the activated dendritic laminae that were blocked by protein synthesis inhibitors.
These results suggested that synaptic activity triggered a local synthesis of
CAMKII protein within the activated dendrites.

5.3.2 Studies in Intact Animals

Using a similar approach, we have demonstrated that high frequency stimulation of the perforant path projections to the dentate gyrus also cause increases
in immunostaining for CAMKII in the activated dendritic laminae (Steward
and Halpain 1999). There were also alterations in the pattern of immunostaining for MAP2, but the nature of the change was different than was the case
for CAMKII. Specifically, the increases in immunostaining for MAP2 occurred
in the laminae on each side of the activated lamina. The changes in immunostaining for MAP2 were diminished but not eliminated by systemic or local
application of protein synthesis inhibitors. Surprisingly, however, the increases
in immunostaining for CAMKII were not affected by inhibiting protein syn-

thesis. Thus, high frequency synaptic activity can cause domain-specific alterations in the molecular composition of dendrites, but only a portion of the change may be attributable to local protein synthesis. The reason for the different effects of protein synthesis inhibitors in the studies by Ouyang et al. vs Steward and Halpain remains to be resolved.

5.3.3 Studies in Synaptoneurosomes

Other evidence indicating synaptic regulation of translation within dendrites comes from studies that use synaptoneurosome preparations (a subcellular fraction containing terminals and dendrites isolated by filtration techniques). Both depolarization and neurotransmitter activation lead to an increase in the proportion of mRNA associated with polysomes, and in the levels of protein synthesis within cell fragments in the preparation (Weiler and Greenough 1991, 1993). The induction of polysome formation and stimulation of protein synthesis seem to be triggered by metabotropic glutamate receptor (mGluR) activation, because antagonists of mGluRs block both phenomena.

An especially provocative line of evidence has come from the studies of fragile-X mental retardation protein (FMRP) in synaptoneurosomes. FMRP is encoded by a gene called FMR1, which is affected in human fragile X syndrome. Recent evidence suggests that the protein either plays some role in the translation of mRNAs at synapses (Feng et al. 1997), or that the protein is amongst the ones that are synthesized in dendrites (Weiler et al. 1997). This evidence has led to the idea that the neuronal dysfunction that is part of fragile X syndrome may result from a disruption of local synthesis of protein at synapses, which in turn would disrupt synaptic function, especially the capability to undergo long-lasting forms of synaptic plasticity (Comery et al. 1997; Weiler et al. 1997).

More recent studies of synaptoneurosomes prepared from frog tectum have revealed a novel mechanism for synaptic regulation of the translation of the mRNA for the alpha subunit of calcium/calmodulin-dependent protein kinase II (CAMKII). There is evidence that the formation of eye-specific projections in the tectum is regulated by activity, and that NMDA receptors play an important role in this process (Sheetz et al. 1997). Based on these findings, and the fact that CAMKII mRNA is present in dendrites, Sheetz et al. (2000) evaluated how NMDA receptor activation of synaptoneurosomes from the tectum modulated CAMKII synthesis. Their results revealed a surprising and complex translation regulation mechanism. NMDA receptor activation did enhance CAMKII synthesis within the sub-cellular fractions (presumably within the synaptoneurosomes). At the same time, however, there was an increase in the phosphorylation of the initiation factor IF2, which would be expected to decrease the rate of polypeptide elongation (slowing overall protein synthesis rate). This apparent paradox can be explained by the fact that studies of mRNA competition in translation assays reveal that decreases in elongation favor the translation of weakly initiated mRNAs. CAMKII is one of the mRNAs for which

initiation is inefficient, and so general decreases in elongation consequent to
IF2 phosphorylation could lead to increases in CAMKII synthesis. Sheetz et al.
went on to provide evidence in support of this idea by showing that low to
moderate concentrations of cycloheximide (which partially inhibited overall
protein synthesis) caused increases in the synthesis of CAMKII at the same
time that overall levels of protein synthesis were diminished. These results thus
provided evidence of regulation of CAMKII mRNA translation via NMDA
receptor activation.

Other recent studies have revealed another novel mechanism for the control
of the translation of CAMKII mRNA at synapses (Wu et al. 1998). An unusual
feature of CAMKII mRNA is that it contains a sequence in it's $3'$ untranslated
region (UTR) that is a consensus sequence for the binding of cytoplasmic
polyadenylation element (CPE). CPEs are known to play a key role in regulating the translation of maternal mRNAs in oocytes. These mRNAs, which are
inherited from the mother, have a selective localization in the oocyte cytoplasm. The proteins encoded by these maternal mRNAs are often transcription
factors that control the synthesis of genes that regulate the initial development
of polarity in the embryo (i.e. in defining dorsal vs ventral, anterior vs posterior, and also defining particular body regions).

Maternal mRNAs are translationally repressed until fertilization. Upon fertilization, translation repression is relieved through the action of the CPE in
the following way. Translationally repressed maternal mRNAs have very short
poly-A tails. At fertilization, CPE is activated and triggers an elongation of the
poly-A tail of the mRNA, causing translation induction. In an interesting
experiment, Wu et al. (1998) demonstrated that the translation of CAMKII
kinase was regulated in a similar way in brain. In particular, NMDA receptor
activation triggered poly-adenylation of the mRNA for CAMII kinase, which
in turn increased the synthesis of CAMKII protein. They further showed that
this activation could be triggered by behavioral experience (light exposure in
animals raised in the dark). This study thus revealed a second mechanism
through which CAMKII protein synthesis could be regulated at the synapse. It
remains to be seen how this mechanism interacts with the mechanism suggested in the experiments by Sheetz et al.

Taken together, this evidence suggests a complex regulation of the translation of mRNAs at synapses. Activation of particular neurotransmitter receptors appears to play a role in regulating translation, and both metabotropic
and NMDA-type receptors have been implicated in this process. It remains
to be determined whether different mRNAs are controlled in different ways,
or whether different control mechanisms are present at different types of
synapses.

6 Conclusion

In this chapter, we have attempted to provide an overview of what is currently
known regarding the mechanisms through which particular mRNAs come to

be localized at synaptic sites on dendrites. Clearly, this is an active field, and a number of important issues remain to be resolved. Indeed, although we know a number of details about this mechanism, we do not yet have a clear understanding of the purpose(s) of this elaborate mechanism. Certainly, the mechanism seems ideally suited for bringing about the complex modifications in synapses that we know to occur in response to activity, but definitive evidence in this regard is so far lacking. Indeed, the best guess at this point is that local protein synthesis at the synapse is a multipurpose mechanism that plays different roles in different neuron types.

Acknowledgements. Supported by NIH NS12333 (O.S.), and NIH MH53603.

References

Ainger K, Avossa D, Morgan F, Hill SJ, Barry C, Barbarese E, Carson JH (1993) Transport and localization of exogenous myelin basic protein mRNA microinjected into oligodendrocytes. J Cell Biol 123:431–441

Ainger K, Avossa D, Diana AS, Barry C, Barbarese E, Carson JH (1997) Transport and localization elements in myelin basic protein mRNA. J Cell Biol 138:1077–1087

Berry FB, Brown IR (1996) CaM I mRNA is localized to apical dendrites during postnatal development of neurons in the rat brain. J Neurosci Res 43:565–575

Benson DL (1997) Dendritic compartmentation of NMDA receptor mRNA in cultured hippocampal neurons. Neuroreport 8:823–828

Berry FB, Brown IR (1996) CaM I mRNA is localized to apical dendrites during postnatal development of neurons in the rat brain. J Neurosci Res 43:565–575

Bian F, Chu T, Schilling K, Oberdick J (1996) Differential mRNA transport and the regulation of protein synthesis: selective sensitivity of Purkinje cell dendritic mRNAs to translational inhibition. Mol Cell Neurosci 7:116–133

Blichenberg A, Schwanke B, Rehbein M, Garner CC, Richter D, Kindler S (1999) Identification of a *cis*-acting dendritic targeting element in MAP2 mRNAs. J Neurosci 19:8818–8829

Burgin KE, Washam MN, Rickling S, Westgate SA, Mobley WC, Kelly PT (1990) In situ hybridization histochemistry of Ca^{++}/calmodulin-dependent protein kinase in developing rat brain. J Neurosci 10:1788–1798

Chicurel ME, Terrian DM, Potter H (1990) Subcellular localization of mRNA: isolation and characterization of mRNA from an enriched preparation of hippocampal dendritic spines. Soc Neurosci Abstr 16:353

Comery TA, Harris JB, Willems PJ, Oostra BA, Irwin SA, Weiler IJ, Greenough WT (1997) Abnormal dendritic spines in fragile X knockout mice: maturation and pruning deficits. Proc Natl Acad Sci USA 94:5401–5404

Davis L, Banker GA, Steward O (1987) Selective dendritic transport of RNA in hippocampal neurons in culture. Nature 330:447–479

Davis L, Burger B, Banker G, Steward O (1990) Dendritic transport: quantitative analysis of the time course of somatodendritic transport of recently synthesized RNA. J Neurosci 10:3056–3058

Dugich MM, Tocco G, Willoughby DA, Najm I, Pasinetti G, Thompson RF, Baudry M, Lapchak PA, Hefti F (1992) BDNF mRNA expression in the developing rat brain following kainic acid-induced seizure activity. Neuron 8:1127–1138

Feig S, Lipton P (1993) Pairing the cholinergic agonist carbachol with patterned Schaffer collateral stimulation initiates protein synthesis in hippocampal CA1 pyramidal cell dendrites via a muscarinic, NMDA-dependent mechanism. J Neurosci 13:1010–1021

Feng Y, Gutekunst C-A, Eberhart DE, Yi H, Warren ST, Hersch SM (1997) Fragile X mental retar-
    dation protein: nucleocytoplasmic shuttling and association with somatodendritic ribosomes.
    J Neurosci 17:1539–1547

Ferrandon D, Elphick L, Nusslein-Volhard C, Johnson DS (1994) Staufen protein associates with
    the 3′UTR of bicoid mRNA to form particles that move in a microtubule-dependent manner.
    Cell:1221–1232

Furuichi T, Simon-Chazottes D, Fujino I, Yamada N, Hasegawa M, Miyawaki A, Yoshikawa S,
    Guenet J-L, Mikoshiba K (1993) Widespread expression of inositol 1,4,5-trisphosphate recep-
    tor type 1 gene (Insp3r1) in the mouse central nervous system. Recept Chann 1:11–24

Garner CC, Tucker RP, Matus A (1988) Selective localization of messenger RNA for cytoskeletal
    protein MAP2 in dendrites. Nature 336:674–677

Herb A, Wisden W, Catania dMV, Marechal D, Dresse A, Seeberg PH (1997) Prominent dendritic
    localization in forebrain neurons of a novel mRNA and its product, dendrin. Mol Cell Neu-
    rosci 8:367–374

Hinds HL, Ashley CT, Sutcliffe JS, Nelson DL, Warren ST, Housman DE, Schalling M (1993) Tissue
    specific expression of FMR-1 provides evidence for a functional role in fragile X syndrome.
    Nat Gen 3:36–43

Kiebler MA, Hemraj I, Verkade P, Kohrmann M, Fortes P, Marion RM, Ortin J, Dotti CG (1999)
    The mammalian staufen protein localizes to the somatodendritic domain of cultured hip-
    pocampal neurons. J Neurosci 19:274–287

Kislauskis EH, Zhu X, Singer RH (1994) Sequences responsible for intracellular localization of ß-
    actin messenger RNA also affect cell phenotype. J Cell Biol 127:441–451

Kleiman R, Banker G, Steward O (1994) Development of subcellular mRNA compartmentation
    in hippocampal neurons in culture. J Neurosci 14:1130–1140

Knowles RB, Kosik KS (1997) Neurotrophin-3 signals redistribute RNA in neurons. Proc Natl Acad
    Sci USA 94:14804–14808

Knowles RB, Sabry JH, Martone ME, Deerinck TJ, Ellisman MH, Bassell GJ, Kosik KS (1996)
    Translocation of RNA granules in living neurons. J Neurosci 16:7812–7820

Kohrmann M, Luo M, Kaether C, DesGroseillers L, Dotti CG, Keibler MA (1999) Microtubule-
    dependent recruitment of staufen-green fluorescent protein into large RNA-containing
    granules and subsequent dendritic transport in living hippocampal neurons. Mol Biol Cell
    10:2945–2953

Laudry CF, Watson JB, Handley VW, Campagnoni AT (1993) Distribution of neuronal and glial
    mRNAs within neuronal cell bodies and processes. Soc Neurosci Abstr 19:1745

Leski ML, Steward O (1996) Synthesis of proteins within dendrites: ionic and neurotransmitter
    modulation of synthesis of particular polypeptides characterized by gel electrophoresis.
    Neurochem Res 21:681–690

Link W, Konietzko G, Kauselmann G, Krug M, Schwanke B, Frey U, Kuhl K (1995) Somatoden-
    dritic expression of an immediate early gene is regulated by synaptic activity. Proc Natl Acad
    Sci USA 92:5734–5738

Lyford G, Yamagata K, Kaufmann W, Barnes C, Sanders L, Copeland N, Gilbert D, Jenkins
    N, Lanahan A, Worley P (1995) Arc, a growth factor and activity-regulated gene, encodes a
    novel cytoskeleton-associated protein that is enriched in neuronal dendrites. Neuron
    14:433–445

MacDonald PM, Struhl G (1988) Cis-acting sequences responsible for anterior localization of
    bicoid mRNA in Drosophila embryos. Nature 336:595–598

Mayford M, Baranes D, Podsypania K, Kandel E (1996) The 3′-untranslated region of CAMKII
    alpha is a cis-acting signal for the localization and translation of mRNA in dendrites. Proc
    Natl Acad Sci USA 93:13250–13255

Miyashiro K, Dichter M, Eberwine J (1994) On the nature and differential distribution of mRNAs
    in hippocampal neurites: implications for neuronal functioning. Proc Natl Acad Sci USA
    91:10800–10804

Mowry KL, Melton DA (1992) Vegetal messenger RNA localization directed by a 34-nt sequence
    element in Xenopus oocytes. Science 255:991–994

Ouyang Y, Kreiman G, Kantor DB, Rosenstein AJ, Schuman EM, Kennedy MB (1998) Tetanic stimulation produces a similar increase in both nonphosphorylated and autophosphorylated CAMKII in dendrites of hippocampal CA1 neurons. Soc Neurosci Abstr 23:798

Palacios-Pru EL, Palacios L, Mendoza RV (1981) Synaptogenetic mechanisms during chick cerebellar cortex development. J Submicros Cytol 13:145–167

Palacios-Pru EL, Miranda-Contreras L, Mendoza RV, Zambrano E (1988) Dendritic RNA and postsynaptic density formation in chick cerebellar synaptogenesis. Neuroscience 24:111–118

Paradies MA, Steward O (1997) Multiple subcellular mRNA distribution patterns in neurons: a nonisotopic in situ hybridization analysis. J Neurobiol 33:473–493

Perlin JB, Gerwin CM, Panchision DM, Vick RS, Jakoi ER, DeLorenzo RJ (1993) Kindling produces long-lasting and selective changes in gene expression in hippocampal neurons. Proc Natl Acad Sci USA 90:1741–1745

Pierce JP, van Leyen K, McCarthy JB (2000) Translocation machinery for synthesis of integral membrane and secretory proteins in dendritic spines. Nat Neurosci 3:311–313

Prakash N, Fehr S, Mohr E, Richter D (1997) Dendritic localization of rat vasopressin mRNA: ultrastructural analysis and mapping of targeting elements. Eur J Neurosci 9:523–532

Racca C, Gardiol A, Triller A (1997) Dendritic and postsynaptic localizations of glycine receptor alpha subunit mRNAs. J Neurosci 17:1691–1700

Rao A, Steward O (1991a) Evidence that protein constituents of postsynaptic membrane specializations are locally synthesized: analysis of proteins synthesized within synaptosomes. J Neurosci 11:2881–2895

Rao A, Steward O (1991b) Synaptosomal RNA: assessment of contamination by glia and comparison with total RNA. Soc Neurosci Abstr 17:379

Severt WL, Biber TUL, Wu X-Q, Hecht NB, DeLorenzo RJ, Jakoi ER (2000) The suppression of testis-brain RNA binding protein and kinesin heavy chain disrupts mRNA sorting in dendrites. J Cell Sci 112:3691–3702

Sheetz AJ, Nairn AC, Constantine-Paton M (1997) N-methyl-D-aspartate receptor activation and visual activity induce elongation factor-2 phosphorylation in amphibian tecta: a role for N-methyl-D-aspartate receptors in controlling protein synthesis. Proc Natl Acad Sci USA 94:14770–14775

Sheetz AJ, Nairn AC, Constantine-Paton M (2000) NMDA receptor-mediated control of protein synthesis at developing synapses. Nat Neurosci 3:211–216.

Spacek J, Hartmann M (1983) Three-dimensional analysis of dendritic spines. I. Quantitative observations related to dendritic spines and synaptic morphology in cerebral and cerebellar cortices. Anat Embryol 167:289–310

Steward O (1983) Polyribosomes at the base of dendritic spines of CNS neurons: their possible role in synapse construction and modification. Cold Spring Harbor Symp Quant Biol 48:745–759

Steward O (1997) mRNA localization in neurons: a multipurpose mechanism. Neuron 18:9–12

Steward O, Falk PM (1986) Protein synthetic machinery at postsynaptic sites during synaptogenesis; a quantitative study of the association between polyribosomes and developing synapses. J Neurosci 6:412–423

Steward O, Fass B (1983) Polyribosomes associated with dendritic spines in the denervated dentate gyrus: evidence for local regulation of protein synthesis during reinnervation. Prog Brain Res 58:131–136

Steward O, Halpain S (1999) Lamina-specific synaptic activation causes domain-specific alterations in dendritic immunostaining for MAP2 and CAM kinase II. J Neurosci 15:7834–7845

Steward O, Levy WB (1982) Preferential localization of polyribosomes under the base of dendritic spines in granule cells of the dentate gyrus. J Neurosci 2:284–291

Steward O, Reeves TM (1988) Protein synthetic machinery beneath postsynaptic sites on CNS neurons: association between polyribosomes and other organelles at the synaptic site. J Neurosci 8:176–184

Steward O, Ribak CE (1986) Polyribosomes associated with synaptic sites on axon initial segments: localization of protein synthetic machinery at inhibitory synapses. J Neurosci 6:3079–3085

Steward O, Singer RH (1997) The intracellular mRNA sorting system: postal zones, zip codes, mail bags and mail boxes. In: Hartford JB, Morris DR (eds) mRNA metabolism and post-transcriptional gene regulation. Wiley-Liss, New York, pp 127–146

Steward O, Worley PF (submitted) Selective targeting of newly-synthesized *Arc* mRNA to synaptic sites is mediated by NMDA receptor activation

Steward O, Pollack A, Rao A (1991) Evidence that protein constituents of postsynaptic membrane specializations are locally synthesized: time course of appearance of recently synthesized proteins in synaptic junctions. J Neurosci Res 30:649–660

Steward O, Wallace CS, Lyford GL, Worley PF (1998) Synaptic activation causes the mRNA for the IEG *Arc* to localize selectively near activated postsynaptic sites on dendrites. Neuron 21:741–751

Tongiorgi E, Righi M, Cattaneo A (1997) Activity-dependent dendritic targeting of BDNF and TrkB mRNAs in hippocampal neurons. J Neurosci 17:9492–9505

Torre ER, Steward O (1996) Protein synthesis within dendrites: distribution of the endoplasmic reticulum and the Golgi apparatus in dendrites of hippocampal neurons in culture. J Neurosci 16:5967–5978

Valentine g, Chakravarty S, Sarvey J, Bramham C, Herkenham M (2000) Fragile X (fmr1) mRNA expression is differentially regulated in two adult models of activity-dependent gene expression. Mol Brain Res 75:337–341

Verity MA, Brown WJ, Cheung M (1980) Isolation of ribosome containing synaptosome subpopulation with active in vitro protein synthesis. J Neurosci Res 5:143–153

Wallace CS, Lyford GL, Worley PF, Steward O (1998) Differential intracellular sorting of immediate early gene mRNAs depends on signals in the mRNA sequence. J Neurosci 18:26–35

Watson JB, Coulter PM, Margulies JE, de Lecea L, Danielson PE, Erlander MG, Sutcliffe JG (1994) G-protein gamma7 subunit is selectively expressed in medium-sized neurons and dendrites of the rat neostriatum. J Neurosci Res 39:108–116

Weiler IJ, Greenough WT (1991) Potassium ion stimulation triggers protein translation in synaptoneurosomal polyribosomes. Mol Cell Neurosci 2:305–314

Weiler IJ, Greenough WT (1993) Metabotropic glutamate receptors trigger postsynaptic protein synthesis. Proc Nat Acad Sci USA 90:7168–7171

Weiler IJ, Irwin SA, Klintsova AY, Spencer CM, Brazelton AD, Miyashiro K, Comery TA, Patel B, Eberwine J, Greenough WT (1997) Fragile X mental retardation protein is translated near synapses in response to neurotransmitter activation. Proc Natl Acad Sci USA 94:5395–5400

Wu L, Wells D, Tay J, Mendis D, Abborr M-A, Barnitt A, Quinlan E, Heynen A, Fallon JR, Richter JD (1998) CPEB-mediated cytoplasmic polyadenylation and the regulation of experience-dependent translation of alpha-CaMKII at synapses. Neuron 21:1129–1139

Yisraeli JK, Melton DA (1988) The maternal mRNA *Vg1* is correctly localized following injection into *Xenopus* oocytes. Nature 336:592–595

Yisraeli JK, Sokol S, Melton DA (1990) A two-step model for the localization of maternal mRNA in *Xenopus* oocytes: involvement of microtubules and microfilaments in the translocation and anchoring of *Vg1* mRNA. Development 108:289–298

Long-Lasting Hippocampal Plasticity: Cellular Model for Memory Consolidation?

J.U. Frey[1]

1 Introduction

The concept of memory consolidation was introduced into the scientific discourse over a century ago (Muller and Pilzecker 1900). Consolidation in its most general meaning refers to the idea that recently formed memories can sometimes be subject to stabilization over time, rendering them less susceptible to disruption by both new information and brain dysfunction. This process of maturation and stabilization manifests itself at different levels of neuronal organization and at multiple time windows. It is methodologically useful to distinguish between local, cellular consolidation and system consolidation (for review, see Dudai 1996). The first refers to processes that take place locally, in individual nodes of a neuronal circuit, i.e., synapses and neurons, during the first hours after learning. The second refers to processes that occur at the circuit level, may involve progressive reorganization of memory traces throughout the brain, and in some cases last weeks or longer. In the simplest of circuits, cellular and system consolidation may be isomorphic. In more complex circuits, system consolidation may involve additional mechanisms which reflect the selective activity of different circuits; a consequence of this selectivity is the potential, after consolidation, to retrieve memory using circuits that are different from those encoding the memory at an earlier stage (Shadmehr and Holcomb 1997; Squire and Zola 1998).

In this chapter, cellular consolidation will be discussed in more detail. Specifically, and in contrast to the definition given above, it will be shown that processes which serve as adequate cellular correlates, such as the induction of hippocampal long-term potentiation, may carry properties of the local and cellular type. It is this facet of consolidation that has benefited most in recent years from the exciting developments at the cutting edge of cellular and molecular biology. Hence, from now on, whenever the term consolidation is used, unless otherwise specified, it refers to events that take place in synapses and individual neurons within the first hours after specific electrical stimulation or training.

[1] Leibniz-Institute for Neurobiology, Department of Neurophysiology, Brenneckestr. 6, PF 1860, 39008 Magdeburg, Germany

Results and Problems in Cell Differentiation, Vol. 34
D. Richter (Ed.): Cell Polarity and Subcellular RNA Localization
© Springer-Verlag Berlin Heidelberg 2001

In spite of the recent developments in understanding neuronal plasticity, which potentially relate to consolidation, many questions about cellular consolidation remain unanswered. These include basic issues such as *how*, *where*, and *when* consolidation takes place in local nodes in the neuronal circuits that encode experience-dependent internal representations (Dudai 1989). We will try to discuss some of these aspects in relation to long-term potentiation, the most prominent cellular model thought to underlie such functions.

2 How?

2.1 Long-Term Potentiation

It has been known for over a quarter of a century that particular patterns of electrical stimulation in the hippocampal formation can lead to alterations in synaptic efficacy. The classic observations of Bliss and Lomo (1973), studying the perforant path input to the dentate gyrus of anesthetized rabbits, now referred to as 'long-term potentiation' or 'LTP', have been replicated in other mammalian species, in vitro as well as in vivo, and in several pathways of the hippocampal formation. Contemporary studies have revealed that potentiation of synaptic efficacy in different pathways can have different physiological properties reflecting distinct underlying mechanisms. The best studied of these has been referred to as 'associative' or 'NMDA receptor-dependent LTP', to distinguish it from other forms of lasting synaptic change such as E-S potentiation, mossy-fiber potentiation, long-term depression, neurotrophin-induced potentiation, etc. While these latter forms of neuronal plasticity are unquestionably important, we shall hereafter discuss only the associative NMDA receptor-dependent form and refer to it, for simplicity, as 'LTP'. We shall distinguish only between different temporal phases of its expression. Numerous reviews of LTP have been published (e.g. Matthies et al. 1990; Bliss and Collingridge 1993; Bear and Malenka 1994; Frey 1997; Morris and Frey 1997; Frey and Morris 1998a) and the following general understanding of its properties, mechanisms and functional significance has emerged.

With respect to its physiological properties, LTP is defined as a rapidly induced, persistent enhancement in synaptic efficacy lasting at least 1 h. Its induction is 'associative', in that weak patterns of stimulation insufficient to induce LTP on their own can nonetheless result in a persistent synaptic enhancement if they occur in association with depolarization of the target neuron(s) onto which the stimulated pathway is afferent. The resulting LTP is also 'input-specific' in that meeting the conditions for induction results in enhanced synaptic efficacy specific to the synaptic terminals of the activated pathway (or, at least, to closely neighboring synapses). As noted many times, these properties of persistence, associativity and input specificity are desirable properties of a physiological mechanism for storing information at synapses. Later, a further property of hippocampal LTP will be described whereby the persistence of an induced change in synaptic efficacy can be extended by other heterosynaptic patterns of neural activity.

Regarding the underlying neural mechanisms, there is a consensus that activation of a subclass of excitatory postsynaptic glutamatergic receptor, the so-called N-methyl-D-aspartate (NMDA) receptor, is an essential first step in LTP induction. The NMDA receptor, now known to be a complex protein consisting of a number of individual subunits, has the intriguing property of being both ligand- and voltage-gated. When activated by glutamate, and at a particular level of postsynaptic depolarization, calcium (Ca^{2+}) enters the dendrite via the NMDA receptor ion-channel where it activates a chain of intracellular events leading to altered synaptic efficacy. Theories about how the resulting change in synaptic efficacy is achieved include activation of Ca^{2+}-dependent enzymes that phosphorylate receptors and trigger gene expression. Some of these biochemical events are responsible for short-lasting changes (often referred to as early LTP), others cause the early change to be stabilized, i.e., made persistent (late LTP).

With respect to the functional significance of LTP, studies have been conducted exploring whether there is any correlation between behavioral learning and the occurrence or persistence of LTP (for recent reviews, see Alkon et al. 1991; Morris and Davis 1994; Barnes 1995; Cain 1997; Jeffery 1997; Shors and Matzel 1997). Inevitably, as LTP was first discovered in the hippocampus, such studies have tended to focus on types of learning broadly held to be 'hippocampus-dependent' (i.e., thought to engage hippocampal activity and/or impaired by hippocampal lesions). Correlations have been observed between the persistence of LTP and how long such types of memory are retained, as well as between the occurrence of particular types of learning and the activation of the various Ca^{2+}-dependent enzymes. Studies using drugs that antagonize the NMDA receptor, or targeted mutations of NMDA receptor sub-units, have also revealed behaviorally selective learning impairments. The interpretation of many of these studies is very controversial as the techniques used to manipulate LTP generally have multiple effects on brain function.

2.2 Synaptic Tagging and the Variable Persistence of Hippocampal Synaptic Plasticity

In describing the physiological properties of LTP, persistence for at least 1 h was identified as the defining property of the phenomenon. Clearly, persistence is a necessary condition for a putative synaptic mechanism of information storage underlying any kind of long-term memory. A prominent characteristic of event-memory is, however, that some events are remembered for a long time, and others only for a short time. In fact, the human capacity to remember the inconsequential events of the day for any length of time is actually quite limited, although we are generally able to remember such events for a few hours. It follows that, if NMDA receptor-dependent plasticity is an essential prerequisite for event memory, factors contributing to the variable persistence of LTP could be of functional significance with respect to the strength or accessibility of memory 'traces'.

The issue of variable temporal persistence is made more complicated by the possibility that events happening closely in time may be part of a single 'episode', where an episode is defined as a sequence of related events. It would clearly make sense for the encoding system to be organized in such a manner that most or all events associated with an episode are recalled together. Spatial context may be one important feature of this 'binding' process because, if events are remembered with respect to where they happen, this could provide a basis for considering them as part of a single episode. A common spatiotemporal context is, of course, likely to be only one of several determinants of this binding process.

These speculations form part of the intellectual context of a new series of experiments on the persistence of protein synthesis-dependent late LTP (Frey and Morris 1997). Our immediate aim was to address the issue of how the input-specificity of late LTP is realized. The early phase of NMDA receptor-dependent LTP lasting less than 3 h (early LTP) can be dissociated from LTP lasting longer (late LTP) using inhibitors of protein-synthesis. However, whether synthesized in the cell-body (arguably the more important site; Frey et al. 1989a), or in dendrites (Feig and Lipton 1993; Steward and Wallace 1995; Torre and Steward 1996; Kuhl and Skehel 1998), the question arises of how the input specificity of late LTP is achieved without elaborate protein trafficking. One way might be via the creation of a short-lasting 'synaptic tag' at each activated synapse at the time of LTP induction. This tag would have the potential to sequester plasticity-related proteins (PPs) to stabilize early LTP at that synapse and so render it long-lasting. In the simplest case, a single strong input to a population of afferent fibers could (1) induce early LTP, (2) set synaptic tags locally in the postsynaptic compartment of each of the activated synapses, and (3) trigger the biochemical cascades that increase the synthesis of PPs. The diffuse travel of these newly synthesized PPs inside the cell's dendrites would result in tag-protein interactions only at previously activated synapses. This hypothesis makes an intriguing prediction. Provided the creation of these tags is independent of protein synthesis, there is no reason why tags set in the presence of drugs that inhibit protein synthesis should not 'hijack' PPs synthesized earlier, and so stabilize any early LTP-induced after protein synthesis has been shut down. Paradoxically, the synaptic tag mechanism for realizing input specificity allows for the possibility that protein synthesis-dependent LTP can be induced during the inhibition of protein synthesis.

Summarizing our findings (Frey and Morris 1997, 1998a, b), we can conclude that input specificity and temporal persistence of LTP must be determined somewhat separately. Whereas input specificity is determined by the local synaptic activation of NMDA receptors, temporal persistence appears to be determined, at least in part, by the history of activation of the neuron. Moreover, weak afferent events that usually only give rise to transient changes in synaptic efficacy can be made to cause lasting changes in neurons in which the synthesis of PPs has previously been upregulated.

2.3 Requirements for Cellular Consolidation

Let us start with the trigger. Experimental protocols used to elicit consolidation in LTP are usually based on increasing the intensity of input. For example, one tetanization is used for early LTP and three spaced tetanizations are required to elicit the late LTP (Huang and Kandel 1994). However, our studies reveal that, first, a single tetanus can lead to late LTP if its stimulation intensity and the number of stimuli per tetanus are sufficiently high. However, the notion that consolidation is a function of the intensity of a single input might well be too simplistic. It is also not in accord with the natural situation, in which much of the information that we consolidate is distinguished from the non-consolidated material by virtue of context and association rather than intensity. We therefore favor the notion that a time-dependent convergence of two or more events is required. Therefore, our second point is that the history of the neuron is important in the determination of stimulus intensity to induce late LTP (Frey and Morris 1997, 1998a, b).

To illustrate the latter possibility, let us take a closer look at mechanisms of LTP in hippocampal area CA1 again. It has long been thought that NMDA-receptor-dependent LTP is a homosynaptic event, i.e., glutamatergic synapses are capable of inducing and maintaining all processes required to enhance the efficacy at the particular synapse. However, activation of ionotropic glutamatergic receptors alone leads only to a short- (<1 h) but not long-term potentiation, provided that the extracellular ion concentrations are not manipulated (Kauer et al. 1988). Further, it has been shown that late LTP in CA1 and dentate gyrus can be prevented or simulated by inhibitors and activators, respectively, of aminergic, opioid, or metabotropic glutamatergic receptors coupled to the cAMP cascade (Dunwiddie et al. 1982; Bliss et al. 1983; Krug et al. 1983; Gribkoff and Ashe 1984; Hopkins and Johnston 1984; Stanton and Sarvey 1985a–c; Bramham et al. 1988, 1997; Buzsaki and Gage 1989; Frey et al. 1989b, 1990, 1991; Bramham and Sarvey 1996; Frey 1997). We therefore suggest that late LTP requires concomitant activation of different transmitter systems (Frey 1997; Frey and Morris 1998a; Matthies et al. 1990). The typical LTP experiment, involving a brief period of high frequency stimulation (or, with intracellular recording, pairing of pre- and postsynaptic activation), overlooks the more likely situation in the behaving organism, where the synaptic population of an individual CA1 cell is likely to have a unique history and change dynamically with time, with potentiation at some sites matched by heterosynaptic and homosynaptic decreases in synaptic efficacy elsewhere. The artificial massive stimulation, involving simultaneous activation of hundreds of fibers, probably activates more than one kind of neurotransmitter input, and it is that cooperative action of inputs that induces the late LTP. There are other potential players which may be involved in initiating the long-term process, such as the time-dependent ambient intraneuronal level of protein synthesis (see Sect. 4 below). The main take-home message is that consolidation is not merely a function of more-of-the-same (i.e., transmitter), but also of coincidence (i.e., external stimuli, internal states, or both).

Having set the process in motion, two main classes of scenarios can be considered for the conversion of short- into long-term synaptic change: (1) the same molecular species and processes that serve the short-term changes also serve the long-term change; (2) some of the molecular species and processes serving the short- also serve the long-term changes, but additional molecular species and processes are recruited with time.

The first scenario is theoretically possible, such as a shift in the kinetics of autocatalytic protein kinase-phosphatase cascades (Crick 1984; Lisman 1985). Some evidence points to the feasibility of such a switch in vivo (Lisman et al. 1997). In this scenario, macromolecular synthesis is not expected to have a causal role in establishing the persistence of traces over time, but rather to have a role in supplying the synapse with resources for enhanced molecular turnover, in expanding synaptic space for future plasticity, and in homeostatic functions. Using a metaphor of the synapse as a motor vehicle, it has the engine required for a very long ride, but will evidently stop running in the absence of a continuous supply of fuel and spare parts.

The second scenario regards modulation of macromolecular synthesis as causally required for retention of memory over time. This may be done locally at or near the synapse, or cell-wide. Adhering to the above metaphor, the non-consolidated synapse does not have an engine, merely a starter, and the combustion engine, powerful enough to travel for years, must be assembled during consolidation from parts which are either manufactured on the spot, or shipped from the main factory, i.e., the nucleus. Even if the factory exists on the spot, supplies still have to be imported from the nucleus. In other words, it is likely that local, transitional events suffice to establish consolidation, but transcription in the nucleus is needed for maintaining supply in response to enhanced demands by the active synapse, for homeostasis, and for promoting synaptic capabilities for subsequent use.

What macromolecules are expected to be involved in each of the above scenarios? If we focus on the synapse as the critical (and most thoroughly investigated) locus of change, both pre- and postsynaptic changes could be entertained. Presynaptic modifications are expected to consist of alterations in the efficacy of transmitter release. These might be induced either by changes in the availability of Ca^{2+} (from external as well as from internal stores; Reyes and Stanton 1996), or directly in the release machinery. There are various ways of realizing such processes. Experience-dependent phosphorylation of ion channels, for which there is experimental evidence, is one of them (Kandel and Schwartz 1982). So is, potentially, a direct modification in the release machinery. An appealing memory-keeping step is persistent activation of the appropriate protein kinase(s) (Schwartz 1993). Postsynaptic mechanisms are expected to involve alterations in receptors for neurotransmitters and their coupled signal-transduction cascades. Examples are AMPA receptors in LTP, modulated by phosphorylation (Barria et al. 1997); here again, persistent activation of the protein kinases is a candidate for the memory-keeping devices. Cam-KII is a prime suspect, but definitely not the only one (Otmakhov et al. 1997). Modulation of kinases and receptors need not be expressed only in altered receptor responsiveness to an incoming stimulus; it can also involve modification in the interfacing of a membrane-associated receptor with the cytoskeleton and intracellular signal-transduction cascades (e.g., Niethammer et al. 1997; Otmakhov et al. 1997).

So what might consolidation be in each of the above scenarios? In the first one, initiation of long-term memory actually coincides with the short-term, and consolidating consists in recruiting supportive mechanisms. In the second scenario, macromolecular synthesis might be required to generate more of the same or new variants or types of proteins which augment and extend the function of the ones modified in the short term. For example, proteases that degrade the inhibitory, regulatory subunit of cAMP-dependent protein kinase (Hedge et al. 1993). In addition, mechanisms not required in the short-term might be called into service, e.g., intra- and extracellular proteolysis for synaptic remodeling, and alterations in intra- and extracellular architecture brought about by members of the large family of cell adhesion molecules (Fields and Itoh 1996; Martin and Kandel 1996). This scenario indeed depicts consolidation as including mechanisms similar, if not identical, to those recruited in growth and differentiation (Davis et al. 1996). However, since most cellular preparations used to investigate experience-dependent modifications in synaptic efficacy use rather strong, non-physiological stimulation protocols, it is likely that they reveal molecular events that occur in vivo only in response to stress and injury. The specificity of the transcriptional modulation in such preparations to trace formation in the behaving animal must therefore be regarded with some caution. Whatever the case, in the long-term, the synapse, whether strengthened by proliferation or not, is still expected to harbor modified channels, enzymes or receptors of the kinds mentioned above (Martin and Kandel 1996). Again, this suggests that overenthusiastic emphasis on the nucleus, an understandable outcome of the impressive success of developmental cell biology, might even lure us away from mechanisms specific to representational change in the circuit, which operate and change within a much faster time scale (Dudai 1997).

3 Where?

The discussion above leads to the inevitable conclusion that consolidation proceeds at multiple sites within the neuronal domain. Consider in this respect the engineering problem that a neuron faces in determining where consolidated changes in neuronal function are to occur. Cortical pyramidal cells have many thousands of excitatory glutamatergic synapses receiving afferent input from a large number of other cells. An individual cell functions, therefore, in numerous distinct but overlapping circuits, and the spatial distribution of synaptic weights on an individual cell will reflect that fact at any one time. It follows that any decision to consolidate or not to consolidate the synaptic strength of individual synapses cannot be taken exclusively in the cell body, for the desired change is to be specific to a subset of inputs onto that cell. What the cell body can do, however, is to effect a change that creates the potential for consolidation, but allows the final decision as to whether and where this is to occur to be determined locally. This solution is attractive, but it requires a potentially complex interaction between local and central mechanisms.

One engineering solution for avoiding this complexity is to allow all the decision-making to be done locally, once the relevant permissive or instructive information is available. Individual synapses could not only have the machinery for changing synaptic efficacy, but also all the machinery for ensuring that this change is persistent in the face of protein turnover. If this was the case, the cell-body would have only the housekeeping role of synthesizing the mRNAs or proteins required at the periphery, but would play no part in determining whether or where local consolidation occurred. Such an arrangement is, however, wasteful for at least two reasons. First, the neuron is an integrative computing element that, in addition to summating excitatory postsynaptic potentials and enabling action-potential propagation (over a fast time scale), could also maintain a record of its own recent history of activation. This history (or indeed the immediate future patterns of activation) could play a part in making decisions about consolidation. To take advantage of this opportunity, aspects of decision-making cannot be at the local elements of connectivity, but must be at either the cell body or, allowing one further level of complexity, at individual dendritic domains. The key concept here is to remove part (but not all) of the decision-making from sites where only limited information is available, to sites where afferent information can accumulate. Second, an exclusively local mechanism of consolidation may be biochemically wasteful. If macromolecules are necessary for consolidation, and similar, if not identical, molecules are required at all synapses, it may make no sense to endow each of the thousands of synapses on a cell with the machinery for achieving persistent change. Thus, irrespective of the specific mechanisms involved, a distributed process has the merits of intraneuronal integration and biochemical economy – subject to the eventual necessity for achieving input specificity.

The reference above to dendritic domains deserves further comment. In many areas of the neocortex and allocortex, inputs are received in particular cortical layers that correspond to specific domains of a pyramidal cell's architecture. This state of affairs allows the possibility that synapse formation and/or synaptic change and consolidation may be required in these domains of the cell, but not others. Under these circumstances, the question arises of the efficiency of having only the options of local or central decision-making; it may be advantageous to traffic cellular machinery to those domains where it will be used most intensively. To be specific, it may be advantageous to have specific types of RNA or protein synthesis in layer IV of the cortex where thalamic inputs are received; or in specific segments of the medial perforant path where entorhinal input to the dentate gyrus terminates (Lynch and Baudry 1987; Lyford et al. 1995; Kuhl and Skehel 1998). It should be recognized, however, that this gain in efficiency cannot be at the expense of the necessity for dual decision-making and the advantages it confers upon the neuron.

If distributed decision-making occurs, it is essential that there is effective coordination between the different members of the neuronal team. This is the idea behind the concept of 'synaptic tagging' (Frey and Morris 1997), whereby local sites at which consolidation is presumed to occur (e.g., individual synapses) can sequester the products of consolidation processing elsewhere. Consolidation at local sites in a neuronal domain (such as a small population of synapses) is therefore determined in a dual fashion – by the availability of macromolecules that may have been synthesized in response to other events and by local postsynaptic 'tags' that sequester or 'hijack' these proteins and so render permissive the 'final common path' of synaptic consolidation. As outlined above, there is an appealing symmetry about these dual-location arrangements – persistent synaptic change can also occur if the weak stimulation on a pathway (that sets a local synaptic tag) precedes the stronger stimulation of a population of neurons on another pathway, provided the interval between the two patterns of stimulation is quite short (<1–2h). The success of both the strong-before-weak and weak-before-strong protocols in inducing persistent change on the weakly stimulated pathway offers strong evidence that consolidation is a consequence of two or more neuronal processes operating at different loci within cells. Each creates the potential for persistent change; neither can cause it alone.

These experiments were conducted using extracellular recording techniques and stimulation of large numbers of afferent fibers and neurons. Striking evidence for synaptic tagging occurring within individual cells has been presented by Martin et al. (1997) using *Aplysia* cell cultures. In a preparation in which one sensory cell was afferent onto two motoneurons, weak facilitatory activation of one synaptic terminal following prior strong activation of the other revealed a lasting presynaptic facilitation at both. This result extends the synaptic tagging in several respects: it establishes that tag-macromolecule interactions occur within single cells, it indicates that this can happen presynaptically (as well as postsynaptically), and it reveals that dual-control of consolidation occurs in both invertebrate and vertebrate nervous systems.

4 When?

A frequent claim is that consolidation involves a series of discrete phases. This way of thinking invites the notion of critical periods at which it is feasible to interfere with specific players in this cascade of events. According to this view, consolidation may be said to have a start and an end point, with each of its component processes having characteristic durations. Several factors will determine the time-course of these constituent elements of the machinery, including the typical half-life of synaptic proteins, the speed with which molecules are transported to and from the nucleus, rates of protein synthesis, and so forth. Knowledge of these time courses, coupled with the assumption that they occur in a prescribed sequence, would enable one to work out the start and end points of consolidation, and thus the duration of critical periods after learning has taken place.

However, if, as discussed previously, decision-making in a neuronal domain is distributed, it seems to us more likely that different triggers will be responsible for different elements of the consolidation machinery. Temporal intersection of the products of these different mechanisms at local sites will presumably be both necessary and sufficient for consolidation to occur, but there may then be no prescribed time-course or critical periods at which interference with consolidation will always be successful. To spell out this quite radical suggestion more precisely, suppose local consolidation involved the utilization of somatically synthesized plasticity-related proteins at sites at which specific receptor proteins (e.g., AMPA receptors) had recently been phosphorylated. If these plasticity proteins were to be synthesized in advance, perhaps in response to neuromodulatory input elsewhere on the cell, there is no a priori reason why consolidation could not occur very rapidly after the local phosphorylation events had taken place. The time-consuming cycle of translocation to and from the nucleus is finessed by virtue of the prior history of activation of the neuron. Thus, a phase of consolidation sometimes described as being a 'late' phase could often occur quite early. In the cases of late LTP, for example, early LTP may still be necessary for it to occur, but the local changes at synapses that characterize a consolidated alteration to receptor proteins may nonetheless occur quite early after induction. This suggestion does, incidentally, have the immediate practical importance that it could enable techniques such as intracellular or patch-clamp electrophysiology to be deployed in the analysis of 'phases' of memory formation that have hitherto seemed beyond its reach.

A final aspect of the 'when' question has to do with whether consolidation ever ends. We suspect not. A typical pedagogical scenario is the supposition that a labile memory trace is formed first and it is then subject to consolidation beginning at time t1 and ending at time t2. Consolidation complete, it can now survive the winds and waves of any brainstorm that the nervous system may throw at it. Some memories may be like this – being dredged up years after they were supposedly forgotten. But everyday reality is likely to be a more dynamic process of storage, consolidation, retrieval, integration with other new information, re-storage, a new cycle of consolidation and so on. This more dynamic perspective sees memory traces assuming a status where they acquire both persistence and resistance to interference, yet somehow maintain the capacity to be re-sculpted and so transformed. There is a mystery embedded in such a state of affairs, but one which our present understanding of how internal representations are retained, reactivated or reconstructed in retrieval, is still ill-equipped to address.

5 Conclusion

Recent evidence suggests that consolidation is not triggered in an abrupt, step-function manner by intensive input, and does not necessarily unfold in a pre-determined, fixed cascade of developmental decisions. Rather, it appears to involve concert decision-making in multiple sites in the neuronal domain, and an intricate interaction between local and central mechanisms within the neuron. In vivo, it may need coincident inputs to start rolling. Its time course depends on the history of the neuron and the circuit, and is not expected to follow rigid time windows and phases. Whereas macromolecular synthesis appears obligatory for the process to proceed, the possibility that it does not play a causal role in altering representational properties and does not contribute to the specificity of the change, but rather fulfills, post-factum, supportive, homeostatic and possibly preparatory functions, should not be ignored.

Consolidation is not only indispensable for some types of memory; it is also a potential window into the functions of memory at large, the processes that subserve these functions, the mechanisms that embody these processes, and the interaction between levels of organization and function in the brain. Recent investigations of the cellular biology of consolidation begin to expose fine distinctions between the specific and the general of the cellular processes and mechanisms that subserve the conversion of precepts into long-lasting internal representations.

Acknowledgements. I would like to thank Richard G. M. Morris and Yadin Dudai for their intellectual and practical contribution to this manuscript, and in particular for their help in formulating my initial rough thoughts.

References

Alkon DL, Amaral DG, Bear MF, Black J, Carew TJ, Cohen NJ, Disterhoft JF, Eichenbaum H, Golski S, Gorman LK, Lynch G, McNaughton BL, Mishkin M, Moyer JRJ, Olds JL, Olton DS, Otto T, Squire LR, Staubli U, Thompson LT, Wible C (1991) Learning and memory. Brain Res Rev 16:193–220

Barnes CA (1995) Involvement of LTP in memory: are we 'searching under the street light'? Neuron 15:751–754

Barria A, Muller D, Derkach V, Griffith LC, Soderling TR (1997) Regulatory phosphorylation of AMPA-type glutamate receptors by CaM-KII during long-term potentiation. Science 276:2042–2045

Bear MF, Malenka RC (1994) Synaptic plasticity: LTP and LTD. Curr Opin Neurobiol 4:389–399

Bliss TV, Lomo T (1973) Long-lasting potentiation of synaptic transmission in the dentate area of the anaesthetized rabbit following stimulation of the perforant path. J Physiol (Lond) 232:331–356

Bliss TVP, Collingridge GL (1993) A synaptic model of memory: long-term potentiation in the hippocampus. Nature 361:31–39

Bliss TVP, Goddard GV, Riives M (1983) Reduction of long-term potentiation in the dentate gyrus of the rat following selective depletion of monoamines. J Physiol (Lond) 334:475–491

Bramham CR, Sarvey JM (1996) Endogenous activation of μ and δ-1 opioid receptors is required for long-term potentiation induction in the lateral perforant path: dependence on GABAergic inhibition. J Neurosci 16:8123–8131

Bramham CR, Errington ML, Bliss TV (1988) Naloxone blocks the induction of long-term potentiation in the lateral but not in the medial perforant pathway in the anesthetized rat. Brain Res 449:352–356

Bramham CR, Bacher-Svendsen K, Sarvey JM (1997) LTP in the lateral perforant path is β-adrenergic receptor-dependent. Neuroreport 8:719–724

Buzsaki G, Gage FH (1989) Absence of long-term potentiation in the subcortically deafferented dentate gyrus. Brain Res 484:94–101

Cain DP (1997) LTP, NMDA, genes and learning. Curr Opin Neurobiol 7:235–242

Crick F (1984) Memory and molecular turnover. Nature 312:101–101

Davis GW, Schuster CM, Goodman CS (1996) Genetic dissection of structural and functional components of synaptic plasticity. III. CREB is necessary for presynaptic functional plasticity (see comments). Neuron 17:669–679

Dudai Y (1989) The neurobiology of memory. Oxford University Press, Oxford

Dudai Y (1996) Consolidation: fragility on the road to the engram. Neuron 17:367–370

Dudai Y (1997) Time to remember. Neuron 18:179–182

Dunwiddie TV, Roberson NL, Worth T (1982) Modulation of long-term potentiation: effects of adrenergic and neuroleptic drugs. Pharmacol Biochem Behav 17:1257–1264

Feig S, Lipton P (1993) Pairing the cholinergic agonist carbachol with patterned Schaffer collateral stimulation initiates protein synthesis in hippocampal CA1 pyramidal cell dendrites via a muscarinic, NMDA-dependent mechanism. J Neurosci 13:1010–1021

Fields RD, Itoh K (1996) Neural cell adhesion molecules in activity-dependent development and synaptic plasticity. Trends Neurosci 19:473–480

Frey U (1997) Cellular mechanisms of long-term potentiation: late maintenance. In: Donahoe JW, Dorsel VP (eds) Neural network models of cognition: biobehavioral foundations. Elsevier, Amsterdam, pp 105–128

Frey U, Morris RGM (1997) Synaptic tagging and long-term potentiation. Nature 385:533–536

Frey U, Morris RGM (1998a) Synaptic tagging: implications for late maintenance of hippocampal long-term potentiation. Trends Neurosci 21:181–188

Frey U, Morris RGM (1998b) Weak before strong: dissociating synaptic tagging and plasticity-factor accounts of late-LTP. Neuropharmacology 37:545–552

Frey U, Krug M, Broedemann R, Reymann K, Matthies H (1989a) Long-term potentiation induced in dendrites separated from rat's CA1 pyramidal somata does not establish a late phase. Neurosci Lett 97:135–139

Frey U, Hartmann S, Matthies H (1989b) Domperidone, an inhibitor of the D2-receptor, blocks a late phase of an electrically induced long-term potentiation in the CA1-region in rats. Biomed Biochim Acta 48:473–476

Frey U, Schroeder H, Matthies H (1990) Dopaminergic antagonists prevent long-term maintenance of posttetanic LTP in the CA1 region of rat hippocampal slices. Brain Res 522:69–75

Frey U, Matthies H, Reymann KG (1991) The effect of dopaminergic D1 receptor blockade during tetanization on the expression of long-term potentiation in the rat CA1 region in vitro. Neurosci Lett 129:111–114

Gribkoff V, Ashe J (1984) Modulation by dopamine of population responses and cell membrane properties of hippocampal CA1 neurons in vitro. Brain Res 292:327–338

Hedge AN, Goldberg AL, Schwartz JH (1993) Regulatory subunits of cAMP-dependent protein kinases are degraded after conjugation to ubiquitin: a molecular mechanism underlying long-term synaptic plasticity. Proc Natl Acad Sci USA 90:7436–7440

Hopkins WF, Johnston D (1984) Frequency-dependent noradrenergic modulation of long-term potentiation in the hippocampus. Science 226:350–352

Huang YY, Kandel ER (1994) Recruitment of long-lasting and protein kinase A-dependent long-term potentiation in the CA1 region of hippocampus requires repeated tetanization. Learning Memory 1:74–82

Jeffery KJ (1997) LTP and spatial learning – where to next? Hippocampus 7:95–110

Kandel ER, Schwartz JH (1982) Molecular biology of learning: modulation of transmitter release. Science 218:433–443

Kauer JA, Malenka RC, Nicoll RA (1988) NMDA application potentiates synaptic transmission in the hippocampus. Nature 334:250–252

Krug M, Chepkova AN, Geyer C, Ott T (1983) Aminergic blockade modulates long-term potentiation in the dentate gyrus of freely moving rats. Brain Res Bull 11:1–6

Kuhl D, Skehel P (1998) Dendritic localization of mRNAs. Curr Opin Neurobiol 8:600–606

Lisman JE (1985) A mechanism for memory storage insensitive to molecular turnover: a bistable autophosphorylating kinase. Proc Natl Acad Sci USA 82:3055–3057

Lisman JE, Malenka RC, Nicoll RA, Malinow R (1997) Neuroscience – learning mechanisms: the case for CaM-KII. Science 276:2001–2002

Lyford GL, Yamagata K, Kaufmann WE, Barnes CA, Sanders LK, Copeland NG, Gilbert DJ, Jenkins NA, Lanahan AA, Worley PF (1995) Arc, a growth factor and activity-regulated gene, encodes a novel cytoskeleton-associated protein that is enriched in neuronal dendrites. Neuron 14:433–445

Lynch G, Baudry M (1987) Brain spectrin, calpain and long-term changes in synaptic efficacy. Brain Res Bull 18:809–815

Martin KC, Kandel ER (1996) Cell adhesion molecules, CREB, and the formation of new synaptic connections. Neuron 17:567–570

Martin KC, Casadio A, Zhu HX, Rose JC, Chen M, Bailey CH, Kandel ER (1997) Synapse-specific, long-form facilitation of *Aplysia* sensory to motor synapses: a function for local protein synthesis in memory storage. Cell 91:927–938

Matthies H, Frey U, Reymann K, Krug M, Jork R, Schroeder H (1990) Different mechanisms and multiple stages of LTP. Adv Exp Med Biol 268:359–368

Morris RGM, Davis M (1994) The role of NMDA receptors in learning and memory. In: Collingridge GL, Watkins JC (eds) The NMDA receptor. Oxford University Press, Oxford, pp 340–374

Morris RGM, Frey U (1997) Hippocampal synaptic plasticity: role in spatial learning or the automatic recording of attended experience? Philos Trans R Soc Lond [Biol] 352:1489–1503

Muller GE, Pilzecker A (1900) Experimentelle Beiträge zur Lehre von Gedächtnis. Z Psychol 1:1–300

Niethammer M, Kim E, Sheng M (1997) Interaction between the C terminus of NMDA receptor subunits and multiple members of the PSD-95 family of membrane associated guanylate kinases. J Neurosci 16:2157–2163

Otmakhov N, Griffith LC, Lisman JE (1997) Postsynaptic inhibitors of calcium/calmodulin-dependent protein kinase type II block induction but not maintenance of pairing-induced long-term potentiation. J Neurosci 17:5357–5365

Reyes M, Stanton PK (1996) Induction of hippocampal long-term depression requires release of Ca^{2+} from separate presynaptic and postsynaptic intracellular stores. J Neurosci 16:5951–5960

Schwartz JH (1993) Cognitive kinases. Proc Natl Acad Sci USA 90:8310–8313

Shadmehr R, Holcomb HH (1997) Neural correlates of motor memory consolidation. Science 277:821–825

Shors TJ, Matzel LD (1997) Long-term potentiation: what's learning got to do with it? Behav Brain Sci 20:597+

Squire LR, Zola SM (1998) Episodic memory, semantic memory, and amnesia. Hippocampus 8:205–211

Stanton PK, Sarvey JM (1985a) The effect of high-frequency electrical stimulation and norepinephrine on cyclic AMP levels in normal versus norepinephrine-depleted rat hippocampal slices. Brain Res 358:343–348

Stanton PK, Sarvey JM (1985b) Depletion of norepinephrine, but not serotonin, reduces long-term potentiation in the dentate gyrus of rat hippocampal slices. J Neurosci 5:2169–2176

Stanton PK, Sarvey JM (1985c) Blockade of norepinephrine-induced long-lasting potentiation in the hippocampal dentate gyrus by an inhibitor of protein synthesis. Brain Res 361:276–283

Steward O, Wallace CS (1995) mRNA distribution within dendrites: relationship to afferent innervation. J Neurobiol 26:447–458

Torre ER, Steward O (1996) Protein synthesis within dendrites: glycosylation of newly synthesized proteins in dendrites of hippocampal neurons in culture. J Neurosci 16:5967–5978

Neuronal RNA Localization and the Cytoskeleton

Gary J. Bassell[1] and Robert H. Singer[2]

The ability of a neuronal process to grow and be properly directed depends upon the growth cone, whose shape and sensory capabilities are influenced by dynamic cytoskeletal filament systems. Microfilaments, which are composed of actin, can rapidly form bundles that affect filopodial protrusions, growth cone motility and process growth. An important area of basic research is to understand how the neuron targets cytoskeletal precursors over considerable distances to reach the growth cone for their utilization in filopodial protrusion formation and control of process outgrowth. The localization of mRNAs to distinct cellular compartments has been shown to be an important protein sorting mechanism in many cell types, which is used to generate cell polarity. It is likely that neurons also utilize mRNA sorting and localized synthesis as a means to influence cytoskeletal organization, neuronal polarity, and process outgrowth. This chapter will summarize recent advances in our understanding of mRNA targeting mechanisms in neurons with an emphasis on localized synthesis of cytoskeletal proteins in growth cones and neuronal processes. We will also discuss the role of cytoskeletal filament systems in the transport and localization of specific mRNAs in neuronal processes.

1 Localization of Cytoskeletal Proteins to Neuronal Processes and Growth Cones

Neurons are quintessential asymmetric cells, having two morphologically and functionally distinct types of processes, axons and dendrites. The development of neuronal shape and polarity involves the extension of neuronal processes from the cell body which must traverse long and complex paths to reach their appropriate targets. Growth cones are specialized motile structures at the termini of these processes that respond to extracellular cues and influence both the rate and direction of process outgrowth. Process elongation and growth cone motility depend upon a constant supply of unpolymerized actin and tubulin to assemble new cytoskeletal polymers that will induce membrane pro-

[1] Department of Neuroscience, Albert Einstein College of Medicine, 1300 Morris Park Avenue, Bronx, New York 10461, USA
[2] Department of Anatomy and Structural Biology, Albert Einstein College of Medicine, 1300 Morris Park Avenue, Bronx, New York 10461, USA

Results and Problems in Cell Differentiation, Vol. 34
D. Richter (Ed.): Cell Polarity and Subcellular RNA Localization
© Springer-Verlag Berlin Heidelberg 2001

trusion and promote process outgrowth (Forscher and Smith 1988; Okabe and Hirokawa 1991; Tanaka and Sabry 1995). Disruption of actin filaments in growth cone filopodia has been shown to result in abnormal pathfinding decisions in vivo (Bentley and Toroian-Raymond 1986). A major challenge of neuronal cell biology is to understand how cytoskeletal precursors are delivered to the growing axon and growth cone. Defects in the ability of the neuron to transport and localize cytoskeletal components to distal locations could have deleterious effects on neuronal polarity and cytoarchitecture and lead to abnormal development or degenerative processes.

Not only must the neuron be able to target cytoskeletal proteins to growth cones, it must also be able to target a distinct set of proteins whose identity, stoichiometry, and structural organization differ from that of the perikarya. Of fundamental interest is identification of how the cytoskeletal composition of the growth cone differs from the perikarya, and the identification of mechanisms involved in this sorting and assembly. One mechanism to provide axons and growth cones with specific cytoskeletal proteins is to actively transport them into processes and growth cones following their synthesis within the cell body. In contrast to the fast transport of membrane-bound organelles, the transport of cytoskeletal proteins and/or complexes occurs by a slow transport mechanism which has not been defined. Some studies have suggested the idea of polymer sliding, where the filaments themselves are transported down the axon (Lasek 1986). Others studies suggest that cytoskeletal proteins are synthesized in the cell body and are transported as monomers or possibly oligomers which then incorporate laterally into axonal filaments (Nixon 1987; Okabe and Hirokawa 1990; Sabry et al. 1995; Takeda and Hirokawa 1995). What also remains controversial is whether a slow transport mechanism from the cell body is sufficient to promote renewal of cytoskeletal proteins and maintain a large axonal mass (Koenig 1991).

It has been proposed that the transport of mRNAs into processes may provide an additional mechanism for the localization of newly synthesized cytoskeletal proteins (Crino and Eberwine 1997; Bassell et al. 1998; Zhang et al. 1999). Localized mRNAs could be translated at their site of localization, providing immediate assembly of structural components where they are needed. An mRNA localization mechanism would be more efficient than having to transport each protein molecule from the perikarya, as each localized mRNA could generate thousands of proteins locally. Contemporary models for cytoskeletal transport have not considered this alternative mechanism (Takeda and Hirokawa 1995; Tanaka et al. 1995). However, evidence continues to emerge which documents the localization and translation of mRNAs within neuronal processes. In situ hybridization analysis revealed that mRNAs for the microtubule associated protein, MAP2, were concentrated within dendrites of rat brain tissue sections (Garner et al. 1988). Similar dendritic mRNA distributions for MAP2 have also been observed in cultured hippocampal (Kleiman et al. 1990) and sympathetic neurons (Bruckenstein et al. 1990). Not all mRNAs encoding cytoskeletal filament proteins are localized to neuronal processes.

For example, tubulin appears to be restricted to the cell body, according to in situ hybridization analysis of rat brain and cultured hippocampal neurons (Garner et al. 1988; Kleiman et al. 1994).

The localization of mRNAs into axons has also been demonstrated, although not all axons may utilize this mechanism, and many neurons may restrict protein synthetic machinery from entering the axon (Van Minnen 1994). Polyribosomes have also been detected in growing axons and growth cones in vivo (Tennyson 1970). The axonal distribution of tau protein (Binder et al. 1985; Peng et al. 1986; Ferreira et al. 1987; Brion et al. 1988) may be achieved by the localization of tau mRNAs to the proximal segment of growing axons (Litman et al. 1993). Tau also appears to be transported down axons by a slow transport mechanism which is post-translational (Mercken et al. 1995). Nonetheless, the delivery of tau mRNA to the axonal compartment may facilitate the transport of tau protein to more distal locations (Litman et al. 1993). Tropomyosin mRNA (Tm-5) extends into the proximal segment of growing axons and correlates with the axonal localization of Tm-5 protein (Hannan et al. 1995).

Other mRNAs may be targeted beyond the proximal segment of the axon. mRNAs encoding neurofilament proteins are localized to invertebrate axons and are associated with axonal polyribosomes morphologically and biochemically (Crispino et al. 1993a,b, 1997). Several reports have documented the presence of actin mRNA and actin protein synthesis within vertebrate and invertebrate axons. Actin was found in a cDNA library from squid axoplasm (Kaplan et al. 1992). Axonal preparations from rat sympathetic neurons were shown to be enriched for β-actin mRNAs, whereas tubulin mRNAs were confined to the cell body, suggesting a sequence-specific sorting mechanism (Olink-Coux and Hollenbeck 1996). The axonal synthesis of actin was demonstrated by analysis of proteins synthesized after radiolabeled amino acid incubation of axonal fractions from rat dorsal and ventral roots from rat spinal cord (Koenig 1989, 1991; Koenig and Adams 1982).

Specific mRNAs and translational components have also been localized within growth cones. Using a micropipette to sever the neurite and remove cytoplasm from growth cones, a heterogeneous population of mRNAs was isolated which included MAP2, along with intermediate filament proteins (Crino and Eberwine 1997). Translation of these mRNAs within growth cones was demonstrated by transfection of mRNA encoding an epitope tag and immunofluorescence localization. Dendritic growth cones have also been shown to contain translational machinery by their incorporation of radiolabeled amino acids into proteins following neurite transection (Davis et al. 1992). Ultrastructural analysis has revealed the presence of polyribosomes in growth cones of developing hippocampal neurons (Deitch and Banker 1993). We have shown that β-actin mRNAs extend into processes and growth cones of developing rat cortical neurons, whereas gamma-actin mRNAs were confined to the cell body (Bassell et al. 1998). We have also shown that β-actin mRNA, ribosomal proteins, and elongation factor 1α are present as granules within growth cones

(Bassell et al. 1998). These growth cones contained polyribosomes which were morphologically identifiable by electron microscopy and probes to β-actin mRNA colocalized with translational components in growth cones (Bassell et al. 1998). Our hypothesis is that β-actin mRNA localization into dendritic and axonal processes provides a mechanism for the local enrichment of newly synthesized β-actin monomers within growth cones (Fig. 1). β-actin may be the preferred actin isoform for rearrangements of the actin cytoskeleton which occur in response to signaling events at the membrane. Locally elevated concentrations of the β-isoform may facilitate de novo nucleation of actin polymerization directly at the plasma membrane in response to physiological signals.

Actin is the most abundant protein within the peripheral region of growth cones, known as the leading edge. It is likely that active actin polymerization within the growth cone is supported by an anterograde flux of actin monomers that preferentially incorporates at the barbed end associated with the membrane (Okabe and Hirokawa 1991). Filopodial protrusion is generally followed by a retrograde flow of F-actin which is disassembled at the rear (Bamburg and Bray 1987; Okabe and Hirokawa 1991). Local synthesis of actin within the growth cone may play a role in actin polymerization, filopodial protrusion, and process outgrowth. In fibroblasts, β-actin mRNA localization has been shown

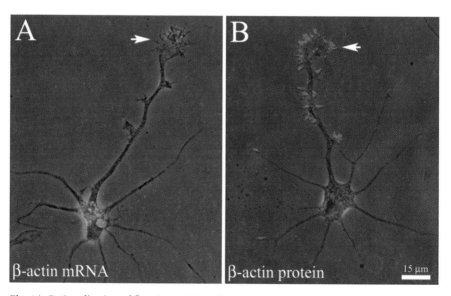

Fig. 1A–B. Localization of β-actin mRNA and protein. **A** Using in situ fluorescence hybridization and digital imaging microscopy, β-actin mRNA was localized in the form of spatially distinct granules, within neuronal processes and growth cones (*arrow*) of cultured chick forebrain neurons (see also Zhang et al. 1999). **B** Immunofluorescence with a monoclonal antibody, which was specific for the β-actin isoform, showed an enrichment of this isoform within growth cones and filopodia (*arrow*)

to enhance cell motility (Kislauskis et al. 1997). Fibroblast lamellae have certain structural and functional similarities to growth cones, hence it is reasonable to suggest that β-actin mRNA targeting also occurs in neuronal processes, yet the cytoskeletal transport mechanisms may differ (Bassell et al. 1998).

There has been an increasing consensus that actin isoforms in several cell types are sorted to different intracellular compartments and that β-actin has a specific role in regions of motile cytoplasm (Herman and D'Amore 1985; Otey et al. 1986; Shuster and Herman 1995; Von Arx et al. 1995; Yao and Forte 1995). β-Actin may be the predominant isoform at submembranous sites where it binds ezrin and thus may be involved in modulation of actin in response to extracellular signals (Shuster and Herman 1995). In cultured forebrain neurons, we have shown that β-actin protein is highly enriched within growth cones and filopodia (Fig. 1B). We have proposed that the asymmetric localization of β-actin protein within neurons is achieved by the localization of β-actin mRNA granules in processes (Fig. 1A) and the subsequent local synthesis of β-actin within growth cones (Bassell et al. 1998; Zhang et al. 1999).

2 Regulation of Neuronal mRNA Localization

The elucidation of regulatory mechanisms for mRNA localization within neuronal growth cones would be important as it would strongly suggest that external cues encountered by the growth cone during pathfinding could directly affect protein synthetic activities locally rather than having to signal similar changes in protein synthesis within the perikarya. The active transport of β-actin mRNA to the fibroblast lamella is induced by serum and platelet-derived growth factor (PDGF); this induction is, in fact, required to obtain maximal rates of cell motility (Latham et al. 1994; Kislauskis et al. 1997). The regulated synthesis of mRNA, its localization, and actin polymerization within neurons could similarly influence process outgrowth. Evidence in support of regulation was suggested by the observation that the amount of mRNA within growth cones was dependent on the stage of development and varied for each mRNA species (Crino and Eberwine 1997). Collapse of growth cones with the calcium ionophore A23187 promoted transport of mRNAs encoding intermediate filaments into growth cones, further suggesting that local synthesis may be regulated (Crino and Eberwine 1997). We have observed that treatment of cultures with db-cAMP, an activator of adenylate cyclase, can induce the transport of β-actin mRNA, but not γ-actin mRNA, into processes and growth cones (Bassell et al. 1998), suggesting that activation of cAMP-dependent protein kinase A could be involved in the regulation of mRNA localization. The neurotrophin, NT-3, was shown to promote the anterograde localization of RNA granules within processes (Knowles and Kosik 1997). Recent work in our laboratory has shown that β-actin mRNAs are localized to growth cones following NT-3 treatment, and that this localization can be blocked by Rp-cAMP, an inhibitor of cAMP-dependent protein kinase A (Figs. 2, 3; see also Zhang et al. 1999). We also observed that NT-3 elicits a transient increase in cyclic APM-

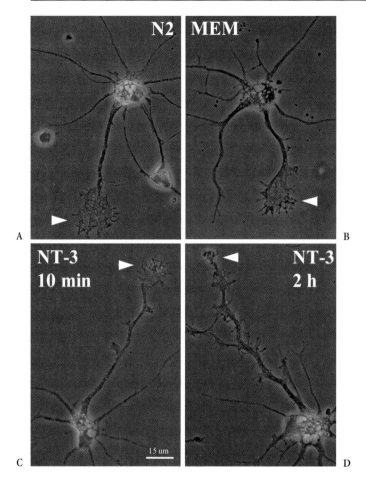

Fig. 2A–D. Signaling of β-actin mRNA localization by neurotrophin-3 (NT-3). **A** In cells cultured under normal conditions, with N2 supplements and astrocyte coculture, β-actin mRNA was prominent in the cell body and localized within the axonal process and growth cone in the form of spatially distinct granules (*arrowhead*). **B** Cells which were starved in minimal essential medium (MEM) for 6 h showed hybridization within the cell body but growth cones did not reveal mRNA granules (*arrowhead*). **C, D** Cells which were starved for 6 h in minimum essential medium (MEM) and then stimulated with NT-3 for 10 min or 2 h were observed to re-localize β-actin mRNA granules within growth cones (*arrow*). *Bar* 15 μm. (Figure modified with permission from Zhang et al. 1999)

dependent protein kinase A (PKA) activity, which precedes the localization of β-actin mRNA (Zhang et al. 1999). An increase in β-actin protein levels and actin polymerization was also observed following NT-3 treatment, which supports our model in which regulated mRNA localization and local synthesis can influence actin organization within growth cones. Further work will be needed to dissect out the downstream targets of trk receptor activation, which influence mRNA localization.

NT-3 increases the density of ß-actin mRNA granules

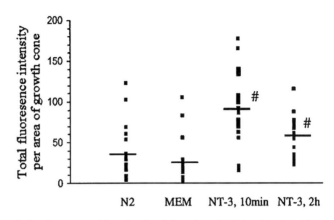

Fig. 3. Quantitative analysis of neurotrophin-stimulated β-actin mRNA localization. Neurons were fixed for in situ hybridization to β-actin mRNA. Twenty growth cones were imaged for each condition with identical exposure times. Data expressed as fluorescence density (total intensity/growth cone area). NT-3 was observed to increase the density of fluorescence signal for β-actin mRNA within growth cones. #, P<0.01 when MEM was compared with N2, or MEM was compared with NT-3, 10 min. or NT-3, 2 h. *N2* N2 normal culture medium, *MEM* starvation in minimum essential medium. (Modified with permission from Zhang et al. 1999)

Table 1. Effects of cytoskeletal perturbation on β-actin mRNA granule density. Fluorescence detection of actin mRNA granules. Data shown are the mean number of granules per 100 μm of process length with standard deviation. For colchicine treatment, the top row indicates randomly selected cells and the bottom row represents those with altered microtubule distribution (74% of cells examined). A Student T-test was performed to measure differences between control and drug-treated cultures (*$P < 0.01$, ** $P < 0.04$, *** $P < 0.005$)

Untreated	37.0 ± 1.4
Cytochalasin-D	*53.0 ± 1.4
Cytochalasin/Actinomysin	38.5 ± 2.6
Colchicine	**23.6 ± 8.4
	***17.0 ± 0.5

3 Mechanism of mRNA Localization

RNA localization is likely to be a multistep process having distinct transport and anchoring components (Bassell and Singer 1999). The interaction of mRNA localization sequences with specific cytoskeletal proteins and filament

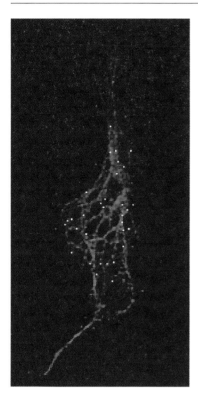

Fig. 4. Colocalization of β-actin mRNA and micro-tubules in growth cones by 3-D digital imag-ing microscopy. To visualize β-actin mRNA and cytoskeletal filaments in cortical neurons with high resolution, a series of optical sections (100 nm) were taken from each cell and further processed using deconvolution algorithms and the application of a point spread function (Bassell et al. 1998). β-Actin mRNA was detected with rhodamine, tubulin protein was detected with fluorescein, and the two processed images were then superimposed. Pixels which con-tained both fluorochromes appeared *white* in optical sections, whereas *red* pixels denote probe that is not within the same pixel as anti-tubulin (*green*). The majority of β-actin mRNA granules colocalized with microtubules (*white pixels*). (Reprinted with permis-sion from Bassell et al. 1998)

systems could provide a mechanism for localizing mRNAs to distinct intra-neuronal regions. Evidence indicates that a nucleic acid-based recognition mechanism exists to sort mRNAs which code for cytoskeletal proteins (Kislauskis and Singer 1992). In many cell types, the presence of specific local-ization sequences within the 3′ untranslated region (UTR) is a common occur-rence. In cultured fibroblasts, a 54 nt (nucleotide) sequence or RNA zip code was found to be necessary and sufficient for localization of β-actin mRNA to the cell periphery. A downstream element, a 43 nt zip code, was shown to have weak localizing activity (Kislauskis et al. 1993, 1994). The targeting of β-actin mRNA to the lamellae is dependent upon microfilaments and not microtubules (Sundell and Singer 1991).

We have shown that the localization of β-actin mRNA in neuronal processes and growth cones is dependent on microtubules (Table 1, Figs. 4, 5). One expla-nation for these cell type-specific differences is that microtubules are the preferred filament system for long-distance translocation, and that β-actin mRNAs can shuttle between both filament systems. In fibroblasts, the *trans*-acting factor, Zip Code Binding Protein, binds the β-actin zip code and is involved in β-actin mRNA localization (Ross et al., 1997). The proteins binding

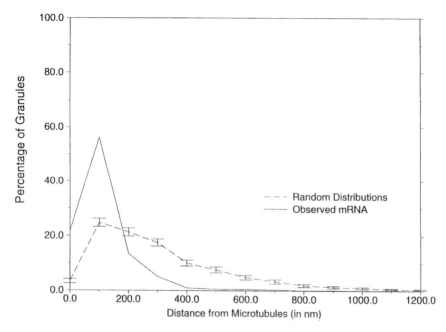

Fig. 5. β-Actin mRNA distribution is significantly nonrandom with respect to microtubules. The distance between β-actin mRNA (brightest voxels) and the nearest tubulin voxel was compared with a randomized distribution. This analysis was performed on a 3-D data set from 100-nm optical sections. The mean and standard deviations of the random distribution are shown. The observed distribution of β-actin mRNA is significantly closer to the microtubules than a random distribution. (Reprinted with permission from Bassell et al. 1998)

Vg1 mRNA in *Xenopus* oocytes were sequenced and their amino acid sequence identified them as a possible orthologue to ZBP (Vera/Vg1 RBP were discovered by different groups but are apparently the same protein) (Deshler et al. 1998; Havin et al. 1998). Thus Vera/Vg1 and ZBP-1 are almost identical proteins operating on different cytoskeletal elements in different cells and species. Further work is needed to determine whether the microtubule-dependent targeting of β-actin mRNA in neurons involves Zip Code Binding Protein and whether the targeting of β-actin mRNA in neurons uses the same *cis*-acting elements as in fibroblasts.

Discrete *cis*-acting localization elements have been identified for polarized cells. In oligodendrocyte processes, a 21-nucleotide sequence has been identified for myelin basic protein mRNA transport into processes (Ainger et al. 1997) which is also dependent on microtubules and kinesin (Carson et al. 1997). RNA localization zip codes have also been identified in oocytes that are involved in targeting proteins to distinct poles, and this sorting plays a fundamental role in the establishment of embryonic polarity (St. Johnston 1995). In *Xenopus* oocytes, Veg1 RNA is localized to the vegetal pole along microtubules, and then is anchored to cortical actin filaments (Yisraeli et al. 1990). These multiple steps may be controlled by distinct *cis*-acting elements within a 340-nucleotide 3'UTR sequence (Mowry and Melton 1992). Two distinct pathways for RNA localization have been identified in *Xenopus*, both of which involve anchoring of RNA to actin filaments within the vegetal cortex, but only one pathway seems to involve microtubule transport of RNA to the cortex (Kloc and Etkin 1995). In *Drosophila* oocytes, the localization of Staufen-bicoid 3'UTR complexes is dependent on microtubules (Ferrandon et al. 1994).

RNA localization sequences have been identified in the translocation of mRNA into neuronal dendrites and axons. The dendritic targeting signal of calcium/calmodulin-dependent protein kinase IIα mRNA is within the 3'UTR and may be required for the localization of CaMKII protein within dendritic spines (Mayford et al. 1996). A *cis*-acting targeting element is required for BC1 RNA localization to neuronal dendrites (Muslimov et al. 1997). The sorting of MAP2 mRNA into dendrites has been observed in hippocampal neurons in culture (Kleiman et al. 1990) and in vivo (Garner et al. 1988). MAP2 localization sequences for the high molecular weight isoform have been suggested to lie within the protein coding sequence (Kindler et al. 1996; Marsden et al. 1996); however, more recent evidence has shown that a 640-nucleotide sequence within the 3'UTR is both necessary and sufficient for dendritic targeting (Blichenberg et al. 1999). A short sequence within the coding region of MAP2 A,B is homologous to the 21-nucleotide RNA transport sequence for myelin basic protein mRNA, and can localize when injected into oligodendrocytes (Munro et al. 1999). It may be that there are redundant signals along the mRNA. The localization of tau mRNAs to the proximal segment of axons requires a 1300-nucleotide sequence within the 3'UTR. The targeting of tau mRNAs to this axonal compartment involves microtubules (Litman et al. 1993) and may be mediated by interactions between 3'UTR sequences and proteins which bind the mRNA to the microtubule (Behar et al. 1995). The mechanism of tau mRNA localization may share certain features with the localization of Vg1 RNA in *Xenopus* oocytes (Litman et al. 1996). Tau mRNA localization sequences injected into oocytes are localized to the vegetal cortex. Tau RNA sequences contain a binding site for Vg1 RNA binding protein and suggest conserved mechanisms of RNA localization between oocytes and neurons. Microtubules may be involved in long-distance translocation of RNA, a mechanism required by these highly polarized cells. Microfilaments may be involved in local movements or mRNA anchoring which could be a common mechanism used by many cell types.

4 mRNA May Be Transported as Particles or Granules

RNAs may be packaged into transport particles or granules which contain multiple mRNA molecules and translational machinery (Bassell et al. 1999). These RNA particles may then be translocated along cytoskeletal filaments via interactions of *cis*-acting elements, motor molecules and accessory proteins. In situ hybridization studies have revealed particulate localization patterns for a variety of mRNAs; these patterns include 'island-like structures' of Xlsirt RNA (Kloc et al. 1993), formation of bicoid RNA 'particles' (Ferrandon et al. 1994), 'granules' of MBP mRNA (Ainger et al. 1993) and granules of β-actin mRNA within neuronal growth cones (Fig. 1; Bassell et al. 1998). The intensity of fluorescence within granules formed by microinjection of MBP RNA labeled with a single fluorochrome suggested that the granules contained multiple mRNA molecules (Ainger et al. 1993). MBP mRNA granules colocalized with arginyl-tRNA synthetase, elongation factor 1α and rRNA, suggesting the presence of a translational unit (Barbarese et al. 1995). MBP RNA granules were estimated to have a radius of between 0.6 and 0.8 μm, suggesting that RNA granules represent a supramolecular complex that could contain several hundred ribosomes (Barbarese et al. 1995). A complex of six proteins has been identified which interact with the RTS, the most abundant being the RNA binding protein, hnRNPA2 (Hoek et al. 1998).

mRNA transport into neuronal processes has been studied using the vital dye SYTO14; mRNA granules were observed which contained ribosomes and elongation factors (Knowles et al. 1996). RNA granules were translocated into processes at a rate of 0.1 μm/s which is similar to rates reported for MBP RNA transport in oligodendrocytes (Ainger et al. 1993). Translocation of RNA granules was blocked by microtubule depolymerization. A subset of RNA granules contained β-actin mRNA, suggesting that the active transport of mRNA may play a role in the targeting of newly synthesized actin into neuronal processes. This mechanism is within the range reported for fast transport (Brady and Lasek 1982) and should be more efficient than the slow transport of newly synthesized actin proteins from the cell body, as each transported mRNA molecule could continuously generate new monomers once localized.

Trans-acting factors involved in localization of specific mRNA granules within neurons have not yet been elucidated. The mammalian homologue of *Drosophila* Staufen has been shown to be. associated with large RNA-containing particles in dendrites (Kiebler et al. 1999). Recently, a Staufen GFP-fusion protein was expressed in cultured hippocampal neurons, allowing analysis of microtubule-dependent particle movements within dendrites (Kohrman et al. 1999).

5 Future Directions

Further work is needed to dissect out the molecular components involved in the assembly of β-actin mRNA granules, their transport along microtubules and translational control within growth cones (Fig. 6). Experiments are in progress to identify *cis*-acting localization elements and RNA-binding proteins which may tether an mRNP complex to microtubule-associated motor molecules. It will be interesting to know whether cell type-specific *cis*-acting elements and/or binding proteins exist for localization of β-actin mRNA over long

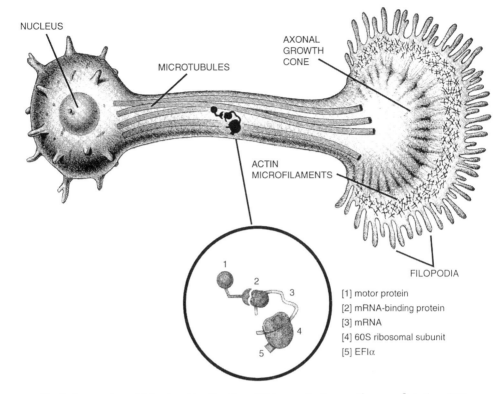

NUCLEUS

AXONAL
GROWTH
CONE

MICROTUBULES

ACTIN
MICROFILAMENTS

FILOPODIA

[1] motor protein
[2] mRNA-binding protein
[3] mRNA
[4] 60S ribosomal subunit
[5] EFIα

Fig. 6. Proposed model for targeting β-actin mRNA granules to growth cones. β-Actin mRNAs are targeted into developing dendrites and axons in the form of granules which contain translational machinery. These granules may contain multiple mRNA molecules, perhaps other mRNAs in addition to β-actin, which encode proteins involved in growth cone dynamics. RNA granules move and anchor to microtubules. Translation of β-actin mRNA may occur preferentially within the growth cone where it can incorporate into filopodial filaments during directed outgrowth. This targeting pathway may be regulated by neurotrophin signaling to influence actin polymerization and directed outgrowth

distances. Microtubules may represent the 'interstate' for mRNA localization over long distances, whereas microfilaments are used for movement of mRNAs over shorter distances. In some cell types or compartments, it is also possible that mRNA granules may switch filament systems. Further work may reveal interactions of both actin and microtubule-based motor molecules at the surface of RNA granules. Once the mechanism for β-actin-mRNA targeting is defined, it will be important to perturb it and assess the consequences on growth cone dynamics and process outgrowth.

There has been seminal work in cell biology which documents the critical involvement of cytoskeletal proteins in growth cone motility and axonal growth. It is also apparent that the activation of intracellular signal transduction pathways by growth factors can have dramatic consequences on the organization of the cytoskeleton, which in turn promotes process outgrowth. Despite the obvious importance of this research and its implications for the development of the nervous system, the mechanism for targeting cytoskeletal machinery in processes remains ill-defined, and it is unclear whether the slow transport of proteins from the cell body is sufficient to respond to the changing needs of the growth cone. An mRNA localization mechanism would be more efficient than having to transport each protein molecule from the perikarya, as each localized mRNA could generate thousands of proteins directly within the growth cone. Localized mRNAs could be translated at their site of localization, providing immediate assembly of structural components where they are needed, at the site of membrane extrusion. RNA granules may represent a novel cytoplasmic organelle capable of localizing specific mRNAs and the necessary translational machinery to maintain an asymmetric distribution of cytoskeletal proteins.

Acknowledgements. We thank our colleagues who contributed to these studies: Honglai Zhang, Andrea Femino, Larry Lifshitz, Ira Herman, and Kenneth Kosik. These studies were supported, in part, by March of Dimes and Muscular Dystrophy Foundation awards (G.J.B.), NSF IBN9811384 (G.J.B.), NIH GM55599 (G.J.B.), and GM 54887 (R.H.S.).

References

Ainger K, Avossa D, Morgan F, Hill SJ, Barry C, Barbarese E, Carson JH (1993) Transport and localization of exogenous MBP mRNA microinjected into oligodendrocytes. J. Cell Biol 123:431–441

Ainger KA, Avossa D, Diana AS, Barry C, Barbarese E, Carson JH (1997) Transport and localization elements in MBP mRNA. J Cell Biol 138:1077–1087

Bamburg JR, Bray D (1987) Distribution and cellular localization of actin depolymerizing factor. J Cell Biol 105:2817–2825

Barbarese E, Koppel DE, Deutscher MP, Smith CL, Ainger K, Morgan F, Carson JH (1995) Protein translation components are colocalized in granules in oligodendrocytes. J Cell Sci 108:2781–2790

Bassell GJ, Zhang HL, Byrd AL, Femino AM, Singer RH, Taneja KL, Lifshitz LM, Herman IM, Kosik KS (1998) Sorting of beta actin mRNA and protein to neurites and growth cones in culture. J Neurosci 18:251–265

Bassell GJ, Oleynikov Y, Singer RH (1999) The travels of mRNA through all cells large and small. FASEB J 13(3):447–454 [see Comment in FASEB J 13(3):419–420]

Behar L, Marx R, Sadot E, Barg J, Ginzburg I (1995) *cis*-Acting signals and *trans*-acting proteins are involved in tau mRNA targeting into neurites of differentiating neuronal cells. Int J Dev Neurosci 13:113–127

Bentley D, Toroian-Raymond A (1986) Disoriented pathfinding by pioneer neurone growth cones deprived of filopodia. Nature 323:712–715

Binder LI, Frankfurter A, Rebhun LI (1985) The distribution of tau in the mammalian central nervous system. J Cell Biol 101:1371–1378

Blichenberg A, Schwanke B, Rehbein M, Garner CC, Richter D, Kindler S (1999) Identification of a *cis*-acting dendritic targeting element in MAP2 mRNAs. J Neurosci 19:8818–8829

Brady ST, Lasek RJ (1982) Axonal transport: a cell biological method for studying proteins that associate with the cytoskeleton. Methods Cell Biol 25:326–398

Brion JP, Guilleminot J, Couchie D, Flament J, Nunez J (1988) Both adult and juvenile tau proteins are axon specific in the developing and adult cerebellum. Neuroscience 25:139–146

Bruckenstein DA, Lein PJ, Higgins D, Fremeau RT (1990) Distinct spatial localization of specific mRNAs in cultured sympathetic neurons. Neuron 5:809–819

Carson JH, Worboys K, Ainger K, Barbarese E (1997) Translocation of MBP mRNA in oligodendrocytes requires microtubules and kinesin. Cell Mot Cyto 38:318–328

Crino PB, Eberwine J (1997) Molecular characterization of the dendritic growth cone. Neuron 17:1173–1187

Crispino M, Capano C, Kaplan B, Guiditta A (1993a) Neurofilament proteins are synthesized in nerve endings in squid brain. J Neurochem 61:1144–1146

Crispino M, Martin R, Giuditta A (1993b) Protein synthesis in a synaptosomal fraction from squid brain. Mol Cell Neurosci 4:366–374

Crispino M, Martin R, Giuditta A (1997) Active polysomes are present in presynaptic endings from squid brain. J Neurosci 17:7694–7702

Davis L, Ping D, DeWitt M, Kater SB (1992) Protein synthesis within neuronal growth cones. J Neurosci 12:4867–4877

Deitch JS, Banker GA (1993) An electron microscopic analysis of hippocampal neurons developing in culture: early stages in the emergence of polarity. J Neurosci 13:4301–4315

Deshler JO, Highett MI, Abramson T, Schnapp BJ (1998) A highly conserved RNA-binding protein for cytoplasmic mRNA localization in vertebrates. Curr Biol 8(9):489–96

Elisha Z, Havin L, Ringel I, Yisraeli JK (1995) Vg1 RNA binding protein mediates the association of Vg1 RNA with microtubules in *Xenopus* oocytes. EMBO J 14:5109–5114

Fay FS, Carrington W, Fogarty KE (1989) Three-dimensional molecular distribution in single cells analyzed using the digital imaging microscope. J Microsc (Oxf) 153:133–149

Ferrandon D, Elphick L, Nusslein-Volhard C, St Johnston D (1994) Staufen protein associates with the 3′UTR of *bicoid* mRNA to form particles that move in a microtubule-dependent manner. Cell 79:1221–1232

Ferreira A, Busciglio J, Caceres A (1987) Microtubule formation and neurite outgrowth in cerebellar macroneurons which develop in vitro: involvement of MAP1a, HMW-MAP2 and tau. Dev Brain Res 34:9–31

Forscher P, Smith SJ (1988) Actions of cytochalasins on the organization of actin filaments and microtubules in a neuronal growth cone. J Cell Biol 107:1505–1516

Garner CC, Tucker RP, Matus A (1988) Selective localization of mRNA for cytoskeletal protein MAP2 in dendrites. Nature 336:674–679

Hannan AJ, Schevzov G, Gunning P, Jeffrey PL, Weinberger RP (1995) Intracellular localization of tropomyosin mRNA and protein is associated with development of neuronal polarity. Mol Cell Neurosci 6:397–412

Havin L, Git A, Elisha Z, Oberman F, Yaniv K, Schwartz SP, Standart N, Yisraeli JK (1998) RNA-binding protein conserved in both microtubule- and microfilament-based RNA localization. Genes Dev 12(11):1593–8

Hoek KS, Kidd GJ, Carson JH, Smith R (1998) hnRNP A2 selectively binds the cytoplasmic transport sequence of myelin basic protein mRNA. Biochemistry 37(19):7021–9

Herman IM, D'Amore PA (1985) Microvascular pericytes contain muscle and nonmuscle actins. J Cell Biol 101:43–52

Hoock TC, Newcomb PM, Herman IM (1991) Beta-actin and its mRNA are localized at the plasma membrane and the regions of moving cytoplasm during the cellular response to injury. J Cell Biol 112:653–664

Kaplan B, Gioio AE, Perrone-Capano C, Crispino M, Giuditta A (1992) β-actin and β-tubulin are components of a heterogeneous mRNA population present in the giant squid axon. Mol Cell Neurosci 3:133–144

Kiebler MA, Hemraj I, Verkade P, Kohrman M, Fortes P, Marion RM, Ortin J, Dotti CG (1999) The mammalian Staufen protein localizes to the somatodendritic domain of cultured hippocampal neurons: implications for its involvement in mRNA transport. J Neurosci 19:288–297

Kindler S, Muller R, Chung WJ, Garner CC (1996) Molecular characterization of dendritically localized transcripts encoding MAP2. Mol Brain Res 36:63–69

Kislauskis EH, Singer RH (1992) Determinants of mRNA localization. Curr Opin Cell Biol 4:975–978

Kislauskis EH, Li Z, Singer RH, Taneja KL (1993) Isoform-specific 3'-untranslated sequences sort α-cardiac and β-cytoplasmic actin messenger RNAs to different cytoplasmic compartments. J Cell Biol 123:165–172

Kislauskis EH, Zhu X, Singer RH (1994) Sequences responsible for intracellular localization of β-actin messenger RNA also affect cell phenotype. J Cell Biol 127:441–451

Kislauskis E, Zhu X, Singer R (1997) β-actin messenger RNA localization and protein synthesis augment cell motility. J Cell Biol 136:1263–1270

Kleiman R, Banker G, Steward O (1990) Differential subcellular localization of particular mRNAs in hippocampal neurons in culture. Neuron 5:821–830

Kleiman R, Banker G, Steward O (1994) Development of subcellular mRNA compartmentation in hippocampal neurons in culture. J Neurosci 14:1130–1140

Kloc M, Etkin LD (1995) Two distinct pathways for the localization of RNAs at the vegetal cortex in *Xenopus* oocytes. Development 121:289–297

Kloc M, Spohr G, Etkin LD (1993) Translocation of repetitive RNA sequences with the germ plasm in *Xenopus* oocyte. Science 262:1712–1714

Knowles RB, Kosik KS (1997) Neurotrophin-3 signals redistribute RNA in neurons. Proc Natl Acad Sci USA 94:14804–14808

Knowles RB, Sabry JH, Martone MA, Ellisman M, Bassell GJ, Kosik KS (1996) Translocation of RNA granules in living neurons. J Neurosci 16:7812–7820

Koenig E (1989) Cycloheximide sensitive methionine labeling of proteins in goldfish retinal ganglion cell axons in vitro. Brain Res 481:119–123

Koenig E (1991) Evaluation of local synthesis of axonal proteins in the goldfish Mauthner cell axon and axons of dorsal and ventral roots of the rat in vitro. Mol Cell Neurosci 2:384–394

Koenig E, Adams P (1982) Local protein synthesizing activity in axonal fields regenerating in vitro. J Neurochem 39:386–400

Kohrman M, Lui M, Kaether C, DesGroseillers L, Dotti CG, Kiebler MA (1999) Microtubule dependent recruitment of Staufen-GFP into large RNA-containing granules and dendritic transport in living hippocampal neurons. Mol Biol Cell 10:2945-2953

Lasek RJ (1986) Polymer sliding in axons. J Cell Sci 5:161–179

Latham VM, Kislauskis EH, Singer RH, Ross AF (1994) Beta-actin mRNA localization is regulated by signal transduction mechanisms. J Cell Biol 126:1211–1219

Litman P, Barg J, Rindzoonski L, Ginzburg I (1993) Subcellular localization of tau mRNA in differentiating neuronal cell culture: implications for neuronal polarity. Neuron 10:627–638

Litman P, Behar L, Elisha Z, Yisraeli JK, Ginzburg I (1996) Exogenous tau RNA is localized in oocytes: possible evidence for evolutionary conservation of localization mechanisms. Dev Biol 176:86–94

Marsden KM, Doll T, Ferralli J, Botteri F, Matus A (1996) Transgenic expression of embryonic MAP2 in adult mouse brain: implications for neuronal polarization. J Neurosci 16:3265–3273

Mayford M, Baranes D, Podsypanina K, Kandel ER (1996) The 3'-untranslated region of CaMKIIa is a *cis*-acting signal for the localization and translation of mRNA in dendrites. Proc Natl Acad Sci USA 93:13250–13255

Mercken M, Fischer I, Kosik KS, Nixon RA (1995) Three distinct axonal transport rates for tau, tubulin and other MAPs. J Neurosci 15:8259–8267

Mowry KL, Melton DA (1992) Veg-1 RNA localization directed by a 340 nt RNA sequence element in *Xenopus* oocytes. Science 255:991–994

Munro TP, Magee RJ, Kidd GJ, Carson JH, Barbarese E, Smith LM, Smith R (1999) Mutational analysis of a heterogeneous nuclear ribonucleoprotein A2 response element for RNA trafficking. J Biol Chem 274:34389–34395

Muslimov IA, Santi I, Perini S, Higgins D, Tiedge H (1997) RNA transport in dendrites: a *cis*-acting signal within BC1 RNA. J Neurosci 17:4722–4733

Nixon RA (1987) The axonal transport of cytoskeletal proteins: a reappraisal. In: Bisby MA, Smith RS (eds) Axonal transport. Liss, New York, pp 175–200

Okabe S, Hirokawa N (1990) Turnover of fluorescently labeled tubulin and actin in the axon. Nature 343:479–482

Okabe S, Hirokawa N (1991) Actin dynamics in growth cones. J Neurosci 11:1918–1929

Olink-Coux M, Hollenbeck PJ (1996) Localization and active transport of mRNA in axons of sympathetic neurons. J Neurosci 16:1346–1358

Otey CA, Lessard JL, Bulinski JC (1986) Immunolocalization of the gamma isoform of nonmuscle actin. J Cell Biol 102:1732–1737

Paves H, Saarma M (1997) Neurotrophins as in vitro guidance molecules for embryonic sensory neurons. Cell Tissue Res 290:285–297

Peng A, Binder L, Black MM (1986) Biochemical and immunological analysis of cytoskeletal domains of neurons. J Cell Biol 102:252–262

Pilkis SJ, Steiner DF, Heinrikson RL (1980) Phosphorylation of rat pyruvate kinase. JBC 255:2770–2775

Ross A, Oleynikov Y, Kislauskis E, Taneja K, Singer R (1997) Characterization of a β-actin mRNA zipcode-binding protein. Mol Cell Biol 17:2158-2165

Sabry J, O'Connor TP, Kirschner MW (1995) Axonal transport of tubulin in Ti1 pioneer neurons. Neuron 14:1247–1256

Shuster CB, Herman IM (1995) Indirect association of ezrin with F-actin:isoform specificity and calcium sensitivity. J Cell Biol 128:837–848

St Johnston D (1995) The intracellular localization of messenger RNAs. Cell 81:161–170

Sundell CL, Singer RH (1991) Requirement of microfilaments in sorting of actin mRNAs. Science 253:1275–1277

Takeda S, Hirokawa N (1995) Tubulin dynamics in neuronal axons. Neuron 14:1257–1264

Tanaka E, Sabry J (1995) Cytoskeletal rearrangements during growth cone guidance. Cell 83:171–176

Tanaka E, Ho T, Kirschner M (1995) The role of microtubule dynamics in growth cone motility and axonal growth. J Cell Biol 128:139–155

Tennyson VM (1970) The fine structure of the axon and growth cone of the dorsal root neuroblast of the rabbit embryo. J Cell Biol 44:62–78

Van Minnen J (1994) RNA in the axonal domain: a new dimension in neuronal functioning? Histochem J 26:377–391

Von Arx P, Bantle S, Soldati T, Perriard J (1995) Dominant negative effect of cytoplasmic actin isoproteins on cardiomyocyte cytoarchitecture and function. J Cell Biol 131:1759–1773

Walsh DA, Van Patten SM (1994) Multiple pathway signal transduction by cAMP dependent protein kinase. FASEB J 8:1227–1236

Yao X, Forte JG (1995) Polarized distribution of actin isoforms in gastric parietal cells. Mol Biol Cell 6:541–557

Yisraeli JK, Sokol S, Melton DA (1990) A two-step model for the localization of maternal mRNA in *Xenopus* oocytes: involvement of microtubules and microfilaments in the translocation and anchoring of Vg1 mRNA. Development 108:289–298

Zhang HL, Singer RH, Bassell GJ (1999) Neurotrophin regulation of beta-actin mRNA and protein localization to neuronal growth cones. J Cell Biol 147:59–70

Transcription Factors in Dendrites: Dendritic Imprinting of the Cellular Nucleus

Jim Eberwine, Christy Job, Janet Estee Kacharmina, Kevin Miyashiro, and Stavros Therianos[1]

1 Introduction

With the advent of new mRNA amplification technologies and improvement in the sensitivity of older methods, an ever-increasing number of mRNAs have been identified in neuronal dendrites. Initially, in situ hybridization was used to localize CamKII (Burgin et al. 1990) and MAP2 (Garner et al. 1988) mRNAs to neuronal dendrites. These experiments were performed by incorporating radioactive nucleotides into cDNA or cRNA probes followed by in situ hybridization and emulsion dipping. Exposure times of a few weeks to a few months were often required to visualize the signal. Because of the limitations of this methodology, only a handful of mRNAs have been identified in dendrites. By combining mRNA amplification methodologies with single dendrite analysis, we have been able to characterize a much larger fraction of the dendritic mRNA population (Miyashiro et al. 1994; Crino and Eberwine 1996). These mRNA amplification methods include aRNA amplification, a linear amplification procedure, which preserves the relative abundances of the amplified mRNAs compared with their original abundances (Van Gelder et al. 1990; Eberwine et al. 1992); and the polymerase chain reaction, which is an exponential amplification procedure (Saikai et al. 1986). Using the aRNA procedure, dendrite cDNA libraries have been made from single dendrites. A photomicrograph of the single dendrite isolation procedure is shown in Fig. 1. Using these methodologies, over 100 mRNAs have been identified in neuronal dendrites, both from primary cultures of neurons as well as from brain sections. These mRNAs fall into multiple functional classes including those encoding cytoplasmic regulatory proteins such as CamKII; integral membrane signaling proteins including glutamate receptors (Miyashiro et al. 1994), GABA receptors (Crino and Eberwine 1996), and Ca^{2+} channels (Crino and Eberwine 1996), plus various cytoskeletal proteins including MAP-2A and ARC (Link et al. 1995; Lyford et al. 1995). While all of these dendritically localized mRNAs are also present in the cytoplasm of neurons, the bulk of cytoplasmically localized mRNAs have not been detected in dendrites.

Dendritically localized mRNAs are thought to be translated in the dendritic compartment. Various types of evidence suggest this, including the electron

[1] Departments of Pharmacology and Psychiatry, University of Pennsylvania, Medical School, 36th and Hamilton Walk, Philadelphia, PA 19104-6084, USA

Results and Problems in Cell Differentiation, Vol. 34
D. Richter (Ed.): Cell Polarity and Subcellular RNA Localization
© Springer-Verlag Berlin Heidelberg 2001

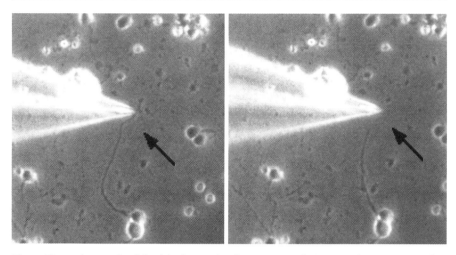

Fig. 1. Photomicrograph of dendrite harvesting for mRNA analysis. A patch pipette is used to sever the dendrite from the cell body. Once severed, the dendrite is aspirated into a micropipette, transferred to a microcentrifuge tube, and mRNA amplification is performed

microscopic discovery of ribosomes in dendrites (Bodian 1965), immunohistochemical localization of various components of the translational machinery in dendrites (Tiedge and Brosius 1996; Gardiol et al. 1999), and metabolic labeling of newly synthesized protein with radiolabeled amino acids in both dendritic and synaptoneurosome preparations (Weiler and Greenough 1993; Torre and Steward 1996). We recently provided a formal molecular proof of dendritic translation of mRNAs by transfecting individual isolated dendrites with mRNAs that contain a sequence encoding a specific epitope tag, c-myc or green fluorescent protein (GFP; Crino and Eberwine 1996; Crino et al. 1998). After transfection of the dendrites, immunohistochemical localization of the epitope tag in the isolated dendrites (which can be detected only if the transfected mRNAs is translated) proved that protein synthesis can occur in dendrites. This assay provides a biochemical model system for characterization of the translational process in neuronal dendrites.

2 CREB mRNA Is Localized to Neuronal Dendrites

Given that dendritically localized mRNAs can be translated, the localization of mRNAs within this subcellular compartment strongly suggests that the raison d'etre for local translation of these mRNAs is the generation of proteins that either (1) serve a function distinct from the same protein in the cell soma or (2) that the dendritically synthesized protein acts locally in the dendrite to modify dendritic function. It is also possible that dendritically translated proteins are different (e.g. post-translational modification) from their soma counterparts and if translocated to the cell soma could function differently from

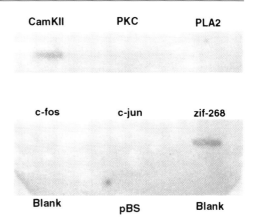

Fig. 2. Mini-expression profile from an individual rat hippocampal neuronal dendrite. These slot blots show the hybridization signal generated with the labeled aRNA population from a single amplified neuronal dendrite. In this study, 1 μm of linearized denatured cDNA containing plasmid is immobilized on the nylon membrane. The blot is prehybridized, hybridized and apposed to film for 72 h with an intensifying screen for signal enhancement. The cDNAs that were analyzed were CamKII, protein kinase C (PKC), phospholipase A2 (PLA2), c-fos, c-jun, and zif-268 with pBluescript (pBS) blotted as a negative hybridization control

the soma-synthesized proteins. This latter idea will be expanded upon for the rest of this review.

   While surveying the population of mRNAs that are present in dendrites, the mRNAs for both cyclic adenosine mono-phosphate responsive binding protein (CREB) and egr-1 (zif-268) were identified (Crino et al. 1998). An expression profile of a single dendrite is shown in Fig. 2 in which CamKII and zif-268 mRNAs are visible while few to no mRNAs for PKC, PLA-2, c-fos and c-jun were detectable. While c-fos mRNA was not detectable in the dendrite that was expression profiled in Fig. 2, c-fos mRNA could occasionally be found in dendrites, with detection in only one out of seven dendrites examined. The finding of transcription factor mRNA localization in dendrites was serendipitous, since it was initially thought that both would serve as negative controls in our search for other dendritically localized mRNAs. While the zif-268 finding has not been described in detail, the discovery of CREB mRNA in this cellular region has implications for our thinking about synaptic communication as shown by long-term potentiation (LTP). In the *Aplysia* and *Drosophila* model systems, CREB has been implicated as a controlling element in generation of LTP as well as various memory functions (Dash et al. 1990). The dogma has been that binding of a ligand to its receptor on the membrane of the post-synaptic dendritic spine would activate a second messenger system, resulting in the production of cAMP that would in turn stimulate the post-translational modification of nuclearly localized CREB, changing its functional activity. While this is still a viable model, with the discovery of CREB mRNA in the dendrite it is possible that CREB protein can be synthesized in the dendrite and then be transported to the nucleus, where it would influence the transcription of various target genes. If this can occur, then direct synthesis of transcription factors in the dendrite and their transport to the nucleus would bypass the integration and averaging of second messenger system signals that would occur in the nucleus if CREB activity is solely activated in the nucleus. We have called this hypothesis 'dendritic imprinting of the nucleus' which is meant to specify a direct connection between generation of transcription factors in the dendrite and their activity in modulating RNA transcription in the cell nucleus (Crino et al. 1998).

Data are accumulating in support of the dendritic imprinting hypothesis. Using CREB as an example; initially CREB mRNA was found in isolated dendrites from primary rat hippocampal pyramidal neurons (Buchhalter and Dichter 1991) using single cell expression profiling methodologies (Crino et al. 1998). This initial observation was confirmed by single-cell PCR, cDNA sequencing and in situ hybridization. Not only was CREB mRNA found in the dendrites of primary cultures of rat hippocampal neurons, but it also was found in dendrites of neurons of the human entorhinal cortex. These data show that the dendritic localization of CREB mRNA crosses species lines and is present at an advanced age (since the human tissue that was examined was from a male over 70 years of age). This suggests a conserved and important role for the dendritic localization of CREB mRNA.

3 CREB Protein Is Present in Dendrites

The presence of mRNA in the dendritic domain does not mean that its corresponding protein is present in the dendrite or that the mRNA can be locally translated. To show that CREB protein was present in neuronal dendrites, immunohistochemistry was performed on cultured rat pyramidal cells using two different antibodies and a blocking control. Results from these studies showed dense nuclear immunoreactivity as well as immunoreactivity in dendrites and dendritic growth cones. To further confirm the identity of the protein as CREB, a new functional in situ assay was developed that we called the in situ southwestern assay (ISSW). Briefly, the DNA sequence encoding the c-AMP response element (CRE) was synthesized on a DNA synthesizer and the oligonucleotide was concatamerized and ^{32}P-labeled. This polynucleotide complex was incubated with paraformaldehyde-fixed primary cell cultures of rat hippocampal neurons. The objective of this analysis was to allow the double-stranded CRE DNA to bind to CREB protein in the dendritic domain of the cultured neurons. Indeed, such localization was observed after emulsion dipping of the cells that had been incubated with the labeled CRE. Various controls were performed that insured the specificity of the ISSW assay.

The presence of CREB protein and mRNA in dendrites does not necessarily mean that the CREB protein was synthesized from the CREB mRNA. It is possible that the protein diffused or was transported from the cell nucleus or soma to the dendrite and that the dendritically localized mRNA is incapable of being translated. To confirm that CREB mRNA can be translated into CREB protein in the dendrite, the single dendrite transfection assay discussed previously was utilized. In this study, a fusion cDNA construct was made that contained the entire CREB protein coding region fused onto a c-myc epitope tag for which a monoclonal antibody was available for detection. mRNA was made from this construct, transfected into individual dendrites that had been severed from their cell bodies and the c-myc epitope was visualized using immunohistochemical procedures to detect the fusion protein. These data directly confirm the hypothesis that CREB mRNA can be translated into protein in the dendritic domain of neurons.

4 CREB Protein Moves from the Dendrite to the Cell Nucleus

If dendritically localized CREB protein plays a role in the regulation of nuclear transcription then this CREB protein must be able to find its way to the cell nucleus. To determine whether dendritically localized CREB can move to the nucleus, fluorescently labeled CREB protein was microperfused into neuronal dendrites and its subcellular localization monitored using fluorescence microscopy. Labeled CREB was observed to move from the site of microperfusion down the dendrite, through the cell body and to concentrate in the cell nucleus. The mechanism of CREB transport is unknown but CREB does contain a nuclear localization signal that facilitates interaction with various intracellular transport proteins such as importin-α. The time course for this transport was quite rapid with much of the transport from the site of perfusion to the nucleus, ~30 µm, complete within 10 min of microperfusion. Further, this dendro-nuclear transport was a specific occurrence, since microperfused fluorescently labeled bovine serum albumin (which is 1.5 times the size of CREB) diffused throughout the cell and was not concentrated in the nucleus. These data show that dendritically localized CREB protein can move from the dendrite to the nucleus.

5 Potential Importance of Dendritic CREB in Modulating Neuronal Functioning

One criticism of the 'dendritic imprinting' hypothesis (that dendritically synthesized and localized transcription factors can translocate to the nucleus to exert a regulatory function on the transcriptional state of a cell) might be that protein entering the nucleus from the dendroplasm would be a 'small amount' compared with the large quantity of protein already present in the nucleus. There are at least three possible scenarios under which this 'small amount' of dendritically translocated transcription factor could exert a transcriptional influence: (1) the transcription factors from the dendrite are actually a large proportion of the cellular transcription factor; (2) a specific phospho-code on the transcription factor (generated by the kinases and phosphatases in the dendrite) might create a functionally distinct form of the transcription factor; (3) the dendritically localized transcription factor complexes with other proteins in the dendrite and the complex translocates into the nucleus where they function as a distinct transcriptional complex to modulate gene expression.

It is reasonable to assume that since a dendrite is small (a tapering diameter averaging 0.5–1 μm) that the amount of dendroplasm is significantly less than that of the cellular cytoplasm. In reality, this turns out not to be true for many neuronal cell types. It is important in discussing dendrite volume to take into account the total dendritic volume, which would be all of the dendrites and their constituent dendritic arbors. For example, in Fig. 3, left-hand panel, an individual biocytin-injected Purkinje cell can be visualized in the context of the Purkinje cell layer of the cerebellum. The Purkinje cell body is indicated by the black arrow while the complex dendritic arbor is highlighted by the white arrows. Even though this is a flat, two-dimensional image, it is clear that the dendritic arborization is quite complex and the dendritic compartment is a significant fraction of the total cellular volume. This is also true of hippocampal pyramidal cells shown in Fig. 3, right-hand panel, where the dendritic compartment (white arrows) is again a large fraction of the total cellular volume. These data suggest that (provided protein synthetic rates are high enough), protein contributions from the dendrites to the cell soma may represent a large component of a protein's total cellular abundance. Clearly, the activation of a single synapse would be unlikely to produce enough protein to affect nuclear CREB functioning. However, the activation of multiple synapses could result in the production of such levels of CREB protein. Alternatively, the generation of a specific form of CREB in the dendrites with a unique function would produce a new type of CREB that could function independently of the bulk of the nuclear CREB protein.

The idea of a specific phospho-code arises from the knowledge that CREB has several potential phosphorylation sites (at least five) embedded within its primary amino acid sequence. Among these sites, phosphorylation of Ser133 is known to modulate CREB-mediated transcription. Phosphorylation of this site was thought to occur exclusively in the nucleus of the neuron (or immediate perinuclear region). However, we have recently demonstrated that CREB-Ser133 can also be phosphorylated in isolated neuronal dendrites in culture. This was accomplished by severing dendrites, stimulating protein synthesis and protein phosphorylation with DHPG, a metabotropic glutamate receptor agonist, and staining the dendrites with an antibody specific for phosphorylated CREB-Ser133. These data provided clear evidence for post-translational modification of CREB directly in dendrites. This does not imply that the same kinase is functioning in both cellular compartments to phosphorylate CREB. It is possible that different kinases phosphorylate Ser133 in the nucleus compared with the dendrite, since kinases can be differentially distributed within the cell. This may be particularly important in defining the activity of proteins that can be phosphorylated, since most kinases are selective and not specific; meaning they will phosphorylate sites other than their preferred sites. For

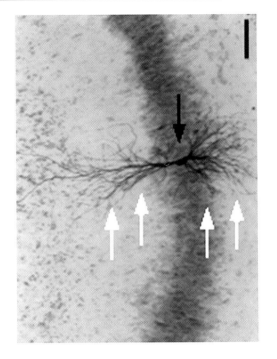

Hippocampal CA1 Pyramidal Cell

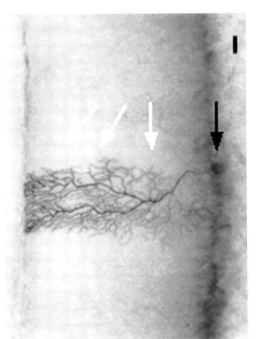

Cerebellar Purkinje Cell

Fig. 3. Dendrite to cell body volume comparison for cerebellar Purkinje and CA1 pyramidal cells. These photomicrographs show biocytin-labeled individual neurons and their dendritic arbors in the Purkinje cell layer of the hippocampus as well as the CA1 pyramidal cells of the hippocampus. The dendritic arbors are quite complex with many branching points extending over a distance of tens to hundreds of microns

example, CamKII is highly enriched in the dendrite while little CamKII exists
in the nucleus. It is unlikely that CamKII phosphorylates the CREB protein that
resides in the nucleus, while in the dendrite it is a distinct possibility. Given
differences in the relative ratios of kinases in the dendrite relative to the
nucleus, it is conceivable that a particular pattern of phosphorylation of the
reactive sites in CREB may result from phosphorylation of CREB in dendrites
which is distinct from that occurring in the nucleus. This different phospho-
pattern could produce different functional states of CREB, such that phosphorylation of sites 1, 2 and 5 may produce CREB of a certain affinity for
hetero-dimerization with other transcription factors while phosphorylation of
sites 2, 3 and 5 may produce a CREB protein with a higher affinity for another
hetero-dimerization partner (schematized in Fig. 4). This hypothesis remains
to be proven (Crino et al. 1998). It should be noted that phosphatases are also
localized to dendrites and we have not yet examined their potential involvement (in conjunction with the kinases) in creating a specific phospho-code on
CREB or other proteins.

Upon micro-perfusion of fluorescently labeled CREB into neuronal dendrites, the fluorescent CREB can be visualized moving to and concentrating
in the cell nucleus. The mechanism by which this vectorial transport occurs is
unclear. In general, proteins that move from the cytoplasm to the nucleus
contain consensus sequences for binding with chaperone molecules that facilitate the transport. CREB contains such a site for a protein called importin
which has been shown to mediate the transport of a variety of proteins into
the nucleus. Whether importin, exists in neuronal dendrites is not yet known.
Regardless, there are other proteins in the dendrite with which CREB can interact, in particular c-fos. Given the co-existence of these proteins in the neuronal
dendrite, it is reasonable to speculate that heterodimers of CREB with zif-268
(and potentially other proteins) can occur in the neuronal dendrite. A priori,
this does not confer any special activity on CREB or zif-268; however, this interaction would likely be mediated by a specific phospho-code generated in the
dendrite for either/both of these transcription factors. This complex, if transported to the nucleus, may be more stable than the nuclear forms of the heterodimer. While such speculation remains unproven, it is highly likely that
CREB interacts with some protein in the dendrite to retain its dendritic localization. This would appear to be the case when one compares the transport of
fluorescently labeled CREB from the dendrite to the nucleus with the presence
of CREB protein that can be normally visualized in dendrites of neurons. The
transport of the labeled CREB was relatively fast, with much of the CREB localizing in the nucleus within 10 min of microperfusion. The observation of CREB
protein in dendrites of normal unstimulated neurons where the protein would
have been there for longer than 10 min suggests that CREB is actively retained
in the dendrite. The identity of such anchoring proteins remains unclear.

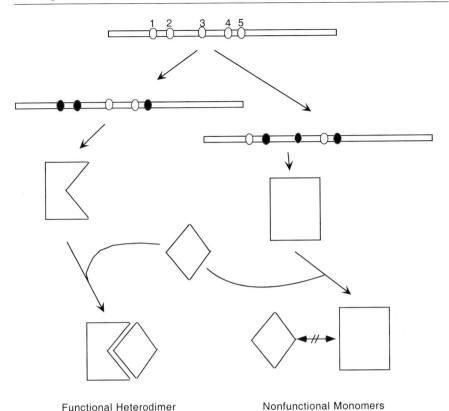

Functional Heterodimer   Nonfunctional Monomers

Fig. 4. Schematic outlining the possible generation of a specific phospho-code for proteins in the dendrite compartment of a cell. This schematic highlights the possible generation of unique phosphorylation patterns on CREB through the phosphorylation of multiple amino acid residues in the protein. The pattern on the *left* (sites 1, 2 and 5 are phosphorylated) produces CREB protein that is able to interact with a specific partner (*diamond* in the center of the schematic). If CREB is phosphorylated as depicted on the *right-hand side* of the schematic (sites 2, 3 and 5 are phosphorylated), then a CREB protein is generated that is incapable of interacting with this specific partner

6 Biological Impetus for Dendritic Imprinting

With the existence of CREB and zif-268 mRNAs in dendrites and the potential role of their cognate proteins in dendritic imprinting, it became of interest to determine what other transcription factors are also localized in dendrites. There are an additional handful of transcription factors including a novel zinc finger mRNA and an mRNA encoding a homeodomain containing protein (Therianos et al., in prep.). There were also many transcription factor mRNAs that were excluded from the dendrite, including c-fos, c-jun and hundreds of mRNAs encoding zinc finger proteins that were screened. These zinc finger-encoding mRNAs were originally isolated in an effort to characterize new transcription factors using the PCR that employed degenerate oligonucleotide

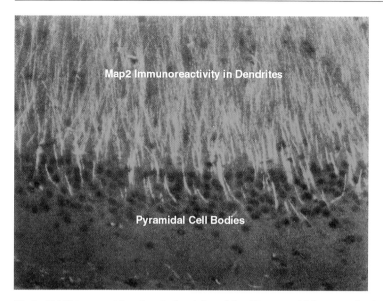

Fig. 5. MAP2 immunohistochemical staining of the CA1 pyramidal neuron dendritic field. MAP2 staining of the dendritic field of CA1 pyramidal neurons shows the dense association of the pyramidal cell dendrites. Understanding the complex interaction of these dendrites with each other, as well as with the axons projecting from the CA3 region of the hippocampus, is still in its infancy

primers directed to the zinc finger region of transcription factor. Since the mRNA for only a few transcription factors has been found in dendrites, there is a clear selectivity in the transport of transcription factor mRNAs, the transport of which identifies these particular mRNAs as a specific subclass of transcription factor mRNAs. The potential involvement of these other transcription factors in dendritic imprinting is just beginning to be appreciated.

The discovery of select transcription factor mRNAs and proteins in dendrites suggests a distinct role for these proteins in regulating neuronal functioning. The dendritic imprinting hypothesis suggests that these transcription factors are produced and modified in response to synaptic activity, are transported to the nucleus and function to alter nuclear transcription. This regulatory process provides an alternative transcriptional regulatory pathway to the previously described nuclear regulation of transcription factor activity. This model provides a framework by which synaptic activity can directly modulate the transcription of responsive genes. This responsiveness is likely to be very specific and exponential in yield. The juxtaposition of nuclear imprinting with the standard nuclear regulation of gene transcription provides an exquisite means by which spatially distinct synaptic responses can exert a profound influence on neuronal gene expression. Indeed, in Fig. 5 the dendritic field of the hippocampal CA1 pyramidal cell layer illustrates the vast opportunity for stimulation of the dendritic inputs of these neurons. It is clear from Fig. 5 that the role of dendritic imprinting in modulating neuronal functioning requires methodologies that permit single dendrite and neuronal cell body functioning to be examined independently of one another. There are many experiments that remain to be done.

Acknowledgements. Brian Johnson and Marc Dichter generously provided the photomicrograph shown in the left panel of Fig. 3. Margie Price and Marc Dichter kindly provided primary hippocampal cell cultures that were used in the described studies. Supported by NIH grants AG09900 and MH58561 to J.E.

References

Bodian D (1965) A suggestive relationship of nerve cell RNA with specific synaptic sites. Proc Natl Acad Sci USA 53:418–425

Buchhalter J, Dichter M (1991) Electrophysiologic comparison of pyramidal and stellate non-pyramidal neurons in dissociated hippocampal cell cultures of rat hippocampus. Brain Res Bull 26:333–338

Burgin K, Waxham M, Rickling S, Westgate S, Mobley W, Kelley P (1990) In situ hybridization histochemistry of Ca^{++}/calmodulin-dependent protein kinase in developing rat brain. J Neurosci 10:1788–1798

Crino P, Eberwine J (1996) Molecular characterization of the dendritic growth cone: regulated mRNA transport and local protein synthesis. Neuron 17:1173–1187

Crino P, Khodakhah K, Becker K, Ginsberg S, Hemby S, Eberwine J (1998) Presence and phosphorylation of transcription factors in dendrites. Proc Natl Acad Sci USA 95:2313–2318

Dash PK, Hochner B, Kandel E (1990) Injection of the cAMP-responsive element into the nucleus of *Aplysia* sensory neurons blocks long-term facilitation. Nature 345:718–721

Eberwine J, Yeh H, Miyashiro K, Cao Y, Nair S, Finnell R, Zettel M, Coleman P (1992) Analysis of gene expression in single live neurons. Proc Natl Acad Sci USA 89:3010–3014

Gardiol A, Racca C, Triller A (1999) Dendritic and postsynaptic protein synthetic machinery. J Neurosci 19:168–179

Garner C, Tucker R, Matus A (1988) Selective localization of messenger RNA for cytoskeletal protein MAP2 in dendrites. Nature 336:374–377

Link W, Konietzko U, Kauselmann G, Krug M, Schwanke B, Frey U, Kuhl D (1995) Somatodendritic expression of an immediate early gene is regulated by synaptic activity. Proc Natl Acad Sci USA 92:5734–5738

Lyford G, Yamagata K, Kaufmann W, Barnes C, Copeland N, Gilbert D, Jenkins N, Lanahan A, Worley P (1995) Arc, a growth factor and activity regulated gene, encodes a novel cytoskeleton-associated protein that is enriched in neuronal dendrites. Neuron 14:433–445

Miyashiro K, Dichter M, Eberwine J (1994) On the nature and distribution of mRNAs in hippocampal neurites: implications for neuronal functioning. Proc Natl Acad Sci USA 91: 10800–10804

Saikai R, Bugawan T, Horn G, Mullis K, Erlich H (1986) Analysis of enzymatically amplified beta-globin and HLA-DQ alpha DNA with allele-specific oligonucleotide probes. Nature 324: 163–166

Tiedge H, Brosius J (1996) Translational machinery in dendrites of hippocampal neurons in culture. J Neurosci 16:7171–7181

Torre E, Steward O (1996) Protein synthesis within dendrites: glycosylation of newly synthesized proteins in dendrites of hippocampal neurons in culture. J Neurosci 16:5967–5978

Van Gelder R, von Zastrow M, Yool A, Dement W, Barchas J, Eberwine J (1990). Amplified RNA (aRNA) synthesized from limited quantities of heterogeneous cDNA. Proc Natl Acad Sci USA 87:1663–1667

Weiler IJ, Greenough W (1993) Metabotropic glutamate receptors trigger postsynaptic protein synthesis. Proc Natl Acad Sci USA 90:7168–7171

Weiler IJ, Wang X, Greenough W (1994) Synapse-activated protein synthesis as a possible mechanism of plastic neural change. Prog Brain Res 100:189–194

RNA Trafficking in Oligodendrocytes

John H. Carson[1], Hongyi Cui[1], Winfried Krueger[1], Boris Schlepchenko[2], Craig Brumwell[3], and Elisa Barbarese[3]

1 Introduction

RNA trafficking provides a mechanism to steer expression of specific proteins to specific compartments of the cell and also to coordinate expression of proteins encoded by RNAs with similar trafficking pathways. Each RNA has an intrinsic intracellular trafficking pathway, determined by *cis*-acting trafficking signals in the RNA and *trans*-acting trafficking factors in the cell. RNA trafficking has been studied to advantage in oligodendrocytes because of their unique morphological features. Oligodendrocytes produce myelin in the central nervous system by extending long, thin dendrites which wrap around axons to form the myelin sheath. This morphology results in topological separation of the different subcellular compartments in the oligodendrocyte, which facilitates resolution of intracellular trafficking intermediates.

The trafficking pathway for myelin basic protein (MBP) mRNA has been extensively characterized in oligodendrocytes. MBP is a major structural component of the myelin sheath and its RNA is transported to the myelin compartment. The pathway by which MBP RNA reaches the myelin compartment has been defined by analyzing the distribution of endogenous MBP mRNA in oligodendrocytes in vivo and by visualizing the movement of exogenous RNA microinjected into cells in culture. Four distinct steps in the MBP RNA trafficking pathway have been delineated: (1) export from the nucleus to the cytoplasm, (2) assembly into granules in the perikaryon, (3) transport along microtubules in the processes, and (4) translation activation in the myelin compartment (Fig. 1). Remarkably, each of these steps appears to be mediated by the same *cis*-acting RNA trafficking signal in MBP RNA, and the same *trans*-acting RNA trafficking factor in the oligodendrocyte (Carson et al. 1998). In this chapter, we describe how these *cis/trans* determinants mediate each step in the MBP RNA trafficking pathway. We also describe a new approach using computational modeling in the Virtual Cell (an object-oriented computational framework for image-based modular spatial modeling of cell physiology;

[1] Department of Biochemistry, University of Connecticut Health Center, Farmington, CT 06030, USA
[2] Department of Physiology, University of Connecticut Health Center, Farmington, CT 06030, USA
[3] Department of Neuroscience, University of Connecticut Health Center, Farmington, CT 06030, USA

Results and Problems in Cell Differentiation, Vol. 34
D. Richter (Ed.): Cell Polarity and Subcellular RNA Localization
© Springer-Verlag Berlin Heidelberg 2001

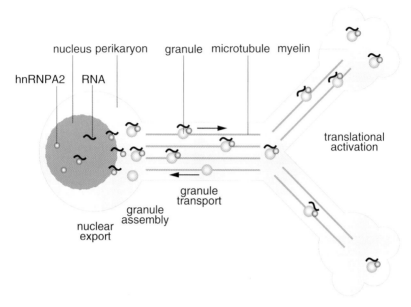

Fig. 1. A2RE/hnRNP A2 mediated RNA trafficking in an oligodendrocyte. A2RE-containing RNA associates with hnRNP A2 in the nucleus. A2RE/hnRNP A2 complexes are exported from the nucleus to the perikaryon where they assemble into granules. Granules are transported in an anterograde direction along microtubules in the processes. At branch points, granules dissociate from the proximal microtubule and reassociate with a distal microtubule. In the myelin compartment translation is activated

Schaff et al. 1997) to analyze how the different components of the RNA trafficking machinery interact to produce the characteristic movement of RNA observed in oligodendrocytes.

2 A2RE Is a *cis*-Acting RNA Trafficking Determinant

The major *cis*-acting determinant for MBP RNA trafficking is a 21-nucleotide sequence, located in the 3'UTR, that is necessary and sufficient for transport of microinjected RNA to the myelin compartment. This sequence was originally termed the RNA transport sequence (RTS; Ainger et al. 1997), but has been renamed the hnRNP A2 response element (A2RE; Munro et al. 1999), because it binds to the *trans*-acting factor, hnRNP A2, which mediates many, if not all, of its RNA trafficking functions. The sequence of the A2RE in MBP RNA is bipartite (Fig. 2), consisting of two partially overlapping, homologous sequences, either of which by itself is sufficient for hnRNP A2 binding and RNA transport. Site-directed mutagenesis has identified specific positions in the A2RE that are essential for hnRNP A2 binding, and other positions that are non-essential (Munro et al. 1999). Mutations that interfere with hnRNP A2

A2RE

$^{5'}$GCCAAGGAGCCAGAGAGCAUG $^{3'}$

hnRNPA2

Fig. 2. A2RE sequence and hnRNP A2 domain structure. In the A2RE sequence, repeated elements are *underlined*. In hnRNP A2, the RBD I and RBD II domains are homologous to RNA binding domains in other hnRNPs, the GRD is a glycine-rich domain and the M9 domain is required for nuclear import and export

binding also inhibit its RNA transport function, while mutations that do not affect binding do not affect transport. This indicates that the RNA transport function of the A2RE requires hnRNP A2 binding.

A2RE-like sequences have been identified in a variety of different RNAs, other than MBP RNA, many of which are known to be transported in other systems (Ainger et al. 1997). One example is myelin oligodendrocytic basic protein (MOBP) RNA, some isoforms of which are transported to the myelin compartment, while others are retained in the perikaryon (Holz et al. 1996; Gould et al. 1999). Transported isoforms of MOBP RNA contain A2RE-like sequences, while non-transported isoforms do not. Two A2RE-like sequences have also been identified in the HIV-1 genome, one in the *gag* gene and one in the *vpr* gene (Mouland et al. 2000). Trafficking of *gag* and *vpr* RNAs has not been examined in HIV infected cells. However, in oligodendrocytes, microinjected HIV RNAs are transported to the myelin compartment, indicating that the A2RE-like sequences in the HIV-1 genome are capable of mediating RNA trafficking. A2RE-like sequences represent a common *cis*-acting element that mediates trafficking of a family of RNAs by a common intracellular trafficking pathway.

3 hnRNP A2 Is a *trans*-Acting RNA Trafficking Factor

The major *trans*-acting factor for A2RE-mediated trafficking is hnRNP A2. Heterogeneous ribonucleoproteins (hnRNPs) comprise a family of proteins implicated in various aspects of RNA metabolism (Dreyfuss et al. 1993). Some hnRNPs, including hnRNP A2, shuttle between the nucleus and the cytoplasm, subserving different functions in each compartment. In other systems, hnRNP A2 orthologues are involved in regulation of splicing (Mayeda et al. 1994), nuclear export (Visa et al. 1996; Daneholt 1997, 1999), RNA localization

(Lall et al. 1999), translation (Hamilton et al. 1999), and RNA degradation (Hamilton et al. 1999). In the A2RE RNA trafficking pathway, hnRNP A2 mediates nuclear export (Brumwell et al. 2000b), granule assembly (Cui et al. 2000a), transport on microtubules (Munro et al. 1999; Brumwell et al. 2000a) and translation activation (Kwon et al. 1999). The domain structure of hnRNP A2 (Fig. 2) consists of two RNA binding domains (RBD 1 and RBD 2), a glycine-rich domain (GRD) and a nuclear import/export domain, referred to as M9. HnRNP A2 is expressed in most tissues, but is particularly abundant in brain (Kamma et al. 1999). When magnetic beads carrying the A2RE sequence are incubated with brain extract, the major protein bound to the beads is hnRNP A2, while two minor bands are isoforms of hnRNP A3 (Hoek et al. 1998). In addition to binding to the A2RE in MBP RNA, hnRNP A2 also binds to an AU-rich element (AURE) in glucose transporter 1 (GLUT1) RNA, which mediates translation repression and RNA stability (Hamilton et al. 1999). HnRNP A2 may represent a new class of 'navigator' proteins that associate with different RNAs in the nucleus and remain associated through subsequent steps in trafficking, mediating different molecular interactions at each step.

4 A2RE/hnRNP A2-Mediated Nuclear Export

HnRNP A2 shuttles between the nucleus and the cytoplasm. Nuclear import is mediated by the M9 domain of hnRNP A2 which interacts with the nuclear import receptor, transportin (Siomi et al. 1997). Nuclear export is also mediated by the M9 domain (Izaurralde et al. 1997), which appears to function as a nuclear import/export switch. In oligodendrocytes, endogenous hnRNP A2 is present in a diffuse distribution in the nucleus and also in granules in the cytoplasm (Brumwell et al. 2000a). The granules in the cytoplasm are believed to represent intracellular trafficking intermediates containing hnRNP A2 associated with endogenous A2RE- or AURE-containing RNAs. Exogenous hnRNP A2, expressed as a green fluorescent protein (GFP) fusion protein, is predominantly retained in the nucleus, with very little in the cytoplasm (Brumwell et al. 2000b). This may indicate that endogenous A2RE- and AURE-containing RNAs in the cytoplasm are already saturated with endogenous hnRNP A2. When excess exogenous A2RE-containing RNA is microinjected into oligodendrocytes, it causes redistribution of exogenous hnRNP A2 from the nucleus to the cytoplasm, suggesting that interaction with A2RE-containing RNA facilitates nuclear export of hnRNP A2. Thus, nuclear-cytoplasmic shuttling of hnRNP A2 appears to be controlled by differential interaction with either transportin, for nuclear import or A2RE, for nuclear export (Fig. 3).

5 A2RE/hnRNP A2-Mediated Granule Assembly

In the cytoplasm of eucaryotic cells, mRNA is assembled into large ribonucleoprotein complexes, termed granules, which can be visualized by light microscopy (Barbarese et al. 1995). Each granule contains multiple mRNA mol-

Fig. 3. Nuclear import/export of hnRNP A2. Cytoplasmic hnRNP A2 associates with transportin and is imported into the nucleus. Nuclear hnRNP A2 associates with A2RE-containing RNA and is exported into the cytoplasm

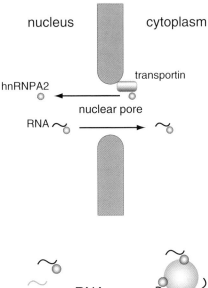

Fig. 4. RNA sorting and granule assembly. RNAs with similar trafficking pathways are co-assembled into the same granule. RNAs with different trafficking pathways are segregated into mutually exclusive granules

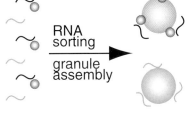

ecules, as well as components of the translation machinery, including amino acyl tRNA synthases, elongation factors and ribosomes. Co-injection experiments with differentially labeled RNAs indicate that RNAs with different trafficking pathways are assembled into different granules, while RNAs with similar trafficking pathways are assembled into the same granules (Cui et al. 2000a; Fig. 4). For example, MBP RNA, which is transported to the myelin compartment, connexin-32 (Cx32) RNA, which is localized to the endoplasmic reticulum, and GFP RNA, which is retained in the perikaryon, are segregated into mutually exclusive granule populations. Whereas HIV-1 gag RNA and vpr RNA, which are both transported to the myelin compartment by virtue of their A2RE-like sequences, are co-assembled into the same granules. Analysis of variance in the ratios of gag and vpr RNAs in individual granules indicates that there are approximately 29 RNA molecules per granule. Co-assembly of RNAs with similar trafficking pathways into the same granule provides a mechanism to ensure co-trafficking of the RNAs and coordinate expression of their encoded proteins in the same compartment of the cell.

RNA sorting and granule assembly are mediated by *cis*-acting RNA trafficking signals in the RNA and by *trans*-acting trafficking factors in the cell. If the A2RE from MBP RNA is inserted into Cx32 RNA or GFP RNA, which normally segregate into mutually exclusive granule populations, the chimeric

RNAs are co-assembled into the same granules as MBP RNA, indicating that the A2RE trafficking signal in MBP RNA mediates RNA sorting, and is epistatic to both the ER trafficking signal(s) in Cx32 RNA and the perikaryon retention signal(s) in GFP RNA. Granules containing A2RE RNAs also contain hnRNP A2, but granules containing nonA2RE RNAs do not. Mutations in the A2RE that interfere with hnRNP A2 binding, also interfere with granule co-assembly and co-localization with hnRNP A2. These results indicate that hnRNP A2 mediates co-assembly of A2RE-containing RNAs into granules. RNA sorting and granule assembly represents the first stage in the trafficking pathway where different RNA molecules are physically segregated from each other.

Cx32 RNA, although it does not contain an A2RE-like sequence and does not co-assemble with MBP RNA, does assemble into granules that contain hnRNP A2, suggesting that it has an hnRNP A2 binding sequence different from the A2RE. Cx32 RNA contains an AU-rich sequence, similar to the AURE in GLUT-1 RNA, which has been shown to bind to hnRNP A2, and to mediate different aspects of RNA trafficking (Hamilton et al. 1999). This implies that there are at least two divergent hnRNP A2-mediated RNA trafficking pathways – A2RE-mediated trafficking to the myelin compartment and AURE-mediated trafficking to the ER.

6 A Dual Motor Mechanism for RNA
Trafficking on Microtubules

In oligodendrocytes, microtubules are oriented with minus ends proximal and plus ends distal (Lunn et al. 1997), which means that anterograde movement of RNA granules requires a plus end motor and retrograde movement requires a minus end motor. Suppressing expression of conventional kinesin with antisense oligonucleotide inhibits A2RE-mediated anterograde transport of RNA from the perikaryon to the myelin compartment, suggesting that the anterograde motor for RNA movement in oligodendrocytes is kinesin (Carson et al. 1997). Injecting antibody to inhibit dynein motor activity causes RNA granules that are normally restricted to the perikaryon to redistribute to the myelin compartment (Cui et al. 2000b), suggesting that dynein functions as a retrograde motor to actively retain certain RNAs in the perikaryon. Co-localization studies indicate that most RNA granules contain both kinesin and dynein molecular motors, implying a dual motor mechanism for RNA trafficking (Cui et al. 2000b).

Different RNAs have different intrinsic trafficking pathways, each requiring characteristic patterns of movement within the cell. For example, MBP RNA is transported to the myelin compartment, requiring long-range anterograde transport, Cx32 RNA is transported to the ER, requiring short-range anterograde transport, and GFP RNA is retained in the perikaryon, requiring retrograde transport. This means that *cis*/*trans* RNA trafficking determinants must differentially regulate the relative activities, or duty cycles, of the kinesin and dynein molecular motors associated with individual granules, in such a way as to determine the direction of RNA movement within the cell.

7 RNA Trafficking in the Virtual Cell

RNA trafficking involves complicated dynamic interactions between RNA granules, molecular motors and microtubules. The various molecular components of the trafficking machinery are known, but it is not clear how they interact to produce the characteristic RNA movements observed in living cells. In order to analyze these interactions systematically, and to identify parameters which are important for determining RNA movement, RNA trafficking was computationally simulated within the framework of the Virtual Cell (an image-based, finite volume approach to modeling cell physiology; Schaff et al. 1997).

In the Virtual Cell, granules are simulated as particles with stochastic behavior. Granules can diffuse throughout the computational domain according to the Gaussian diffusion law (the value of the diffusion coefficient is specified by the user), and undergo elastic collisions at the borders of the domain. Granules can also interact with continuously distributed species, as well as with discrete species. Microtubules are simulated as one-dimensional, oriented contours, whose shapes are defined arbitrarily. Each contour is decomposed into multiple contour elements. Thus, the capture of a granule to a microtubule is reduced to the interaction between a granule and a contour element. In order for a granule to bind to a contour, at least one contour element must be located within a certain radius of the granule (the capture radius), corresponding to the actual size of an RNA granule. Granules can be captured by contours through either anterograde or retrograde motors and, once captured, move along the contour in a direction and with a velocity characteristic of that motor (Fig. 5). Capture rates, release rates, and velocities are independent variables for each motor. Since kinesin is the anterograde motor, and dynein is the retrograde motor for RNA trafficking, the capture rates for anterograde and retrograde movement correspond to the microtubule binding rates for kinesin and dynein, respectively. Since the number of motors per granule may be greater than one, release rates may be significantly lower than capture rates. Release rates for kinesin are generally set lower than for dynein in order to reflect the greater processivity of the kinesin motor relative to dynein. When a granule reaches the end of a contour it immediately falls off and can diffuse in the volume.

The Monte-Carlo technique is used to simulate the stochastic events of granule capture and release. Bayes theorems are used to determine which type of motor binds to the contour at each time step, as follows. If A is the event of granule capture by the contour element within a small time interval, Δt, and H_i ($i = 1,2$) is the event that the particle will try to bind with the i-th motor, then the probability of A can be evaluated as:

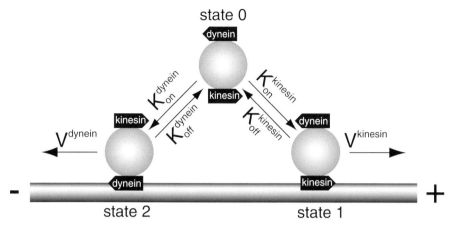

Fig. 5. Stochastic dual motor mechanism for transport of granules on microtubules. Unbound RNA granules (state 0) can be captured by microtubules via either anterograde (state 1) or retrograde (state 2) motors. Each motor has a characteristic capture rate, release rate, and velocity

$$P(A) = \sum_{i=1}^{2} P(H_i)P(A|H_i)$$

where $P(H_i)$ depends on the population of motors on the granule [in our simulations we assume $P(H_1) = P(H_2) = 0.5$] and $P(A|H_i)$ are the corresponding motor on-rates multiplied by Δt. P(A) is the probability that the particle will bind within each time step. If it binds, the conditional probability $P(H_i|A)$ that it will bind by the i-th motor is evaluated as:

$$P(H_i|A) = P(H_i)P(A|H_i)/P(A)$$

The Monte-Carlo approach is used to determine which of the motors binds to the contour element. The granule changes from state '0' in the volume, to state '1' when bound to the contour by the anterograde motor, or state '2' when bound by the retrograde motors. When the granule reaches the end of a microtubule it falls off the contour and changes back to state '0'.

Each of the RNA trafficking functions (diffusion, capture, transport, release) was tested separately, and in combination with the other functions. Where possible, simulation results were compared with either theoretical calculations or experimental results. Here we describe the results of two simulations in the Virtual Cell. In the first simulation, the effect of taxol on RNA trafficking is modeled. In the second simulation, the saltatory movement of RNA granules at branch points, is modeled.

8 The Taxol Paradox

Movement of RNA granules in living cells requires intact microtubules. Taxol stabilizes intact microtubules, and also stimulates polymerization of tubulin monomers into short, randomly-oriented, supernumerary microtubules. Para-doxically, treating oligodendrocytes with taxol inhibits intracellular movement of RNA (Carson et al. 1997). To investigate the taxol paradox in the Virtual Cell, granule movement was simulated in the absence, and in the presence, of excess, randomly oriented contours, and the results were compared with the experimentally observed movement of granules in untreated and taxol-treated cells.

RNA granules were visualized in living cells by labeling with the cell-permeant RNA specific fluorescent dye, SYTO 12 (Molecular Probes, Inc) (Knowles et al. 1996). Time-lapse imaging of SYTO 12-labeled RNA granules in oligodendrocytes distinguished two populations with distinct mobility properties – granules with small maximal displacements (essentially immo-bile) and granules with large maximal displacements (mobile). In untreated cells, the size of the mobile fraction was approximately equal to the size of the immobile fraction. In taxol-treated cells, the size of the mobile fraction of granules was substantially decreased, with a concomitant increase in the immobile fraction.

To simulate granule movement in the virtual cell, granules were introduced close to the minus end of a bundle of ten long parallel contours oriented hor-izontally in the computation volume. To simulate the effect of taxol, an equiv-alent concentration of short, randomly oriented contours were added to the long, oriented contours. Simulations were performed with, and without, the excess short contours, and the proportions of mobile and immobile granules were determined. In the simulation without excess contours, the relative pro-portions of mobile and immobile granules were approximately equal, similar to the experimentally determined proportions in untreated oligodendrocytes. If the parameters of capture rate, release rate and velocity were kept constant, addition of excess contours had little effect on the proportions of mobile and immobile granules, indicating that the mere presence of excess microtubules is not sufficient to inhibit RNA trafficking. However, microtubule concentra-tion is known to affect the ATPase activity of dynein, in vitro (Omoto and Johnson 1986). To simulate this phenomenon in the virtual cell, the capture rate for the retrograde motor was increased, to reflect increased dynein ATPase activity in the presence of excess short microtubules. Under these conditions, the proportion of mobile granules decreased and the proportion of immobile granules increased, similar to the proportions measured experimentally in taxol-treated cells. These results provide an explanation for the taxol paradox – inhibition of RNA movement by taxol is caused by the effect of increased microtubule concentration on the dynein capture rate.

9 Negotiating Branch Points

Electron micrographs of oligodendrocytes reveal that microtubules are dis-continuous at branch points (Lunn et al. 1997). This means that for a granule to negotiate a branch point in the anterograde direction, it must first dissociate from the plus end of a proximal microtubule and then reassociate, via an anterograde motor, with a distal microtubule. Reassociation with a proximal microtubule via either retrograde or anterograde motors or reassociation with a distal microtubule via a retrograde motor will lead to pausing or oscillation. Thus, the behavior of a granule at a branch point may be sensitive to the capture rates, release rates or velocities of both the anterograde and retrograde motors.

Time-lapse imaging of granule movement in living oligodendrocytes indicates a saltatory pattern of movement – with intervals of sustained vectorial movement in individual processes, interrupted by pausing or oscillation at branch points. The duration of the intervals of sustained motion is comparable to the duration of the pauses. The Virtual Cell was used to test whether this behavior could be simulated using a stochastic dual motor model. Multiple contours, representing individual microtubules, were juxtaposed in a geometry reflecting the organization of microtubules at branch points, based on electron microscopic images of oligodendrocytes. Individual granules were introduced at the minus ends of the proximal microtubule bundle and their trajectories were plotted as they negotiated several branch points during the simulation. Parameter sensitivity analysis was used to determine which parameters affected the relative durations of sustained movement and pausing. Capture rates, release rates and velocities for both the anterograde and retrograde motors were varied systematically and independently, and the effects on granule behavior at the branch points were recorded and compared with the behavior observed at branch points in living oligodendrocytes. The results indicate that the duration of granule pausing and oscillation at a branch point is relatively insensitive to capture and release rates for the anterograde and retrograde motors, but is sensitive to the relative velocities of the two motors. When the velocities for the two motors were approximately equal, the relative durations of sustained movement and pausing and oscillation at branch points were comparable to the behavior observed at branch points in living cells. This indicates that the characteristic saltatory motion of RNA granules at branch points in oligodendrocyte processes can be accurately simulated in the Virtual Cell, using image-based microtubule geometries and a stochastic dual motor mechanism with realistic capture, release and velocity parameters for each motor.

Oscillation of granules at branch points may represent a type of exploratory behavior as the granule decides which branch to take. In living cells, granules often tend to move down one branch preferentially. This implies a steering mechanism to bias granule movement towards the preferred branch and/or away from the disfavored branch. In oligodendrocytes, branch point steering may reflect different stages of myelination in each branch. In neurons, branch point steering is believed to reflect differential synaptic activity in the two branches. The nature of the steering mechanism is not known. However, if granule oscillation at branch points reflects multiple cycles of stochastic capture and release by anterograde and retrograde motors, as the Virtual Cell experiments suggest, steering could be accomplished through differential capture rates, release rates or velocities for one, or both, motors. Experiments are in progress to investigate the steering mechanism in living oligodendrocytes, and in the Virtual Cell.

10 Translation Regulation

RNA trafficking steers specific RNAs towards certain subcellular compartments and away from others, thereby maximizing expression of the encoded protein at the target location and minimizing ectopic expression elsewhere in the cell. Implicit in this explanation is the assumption that translation is repressed while the RNA is in transit and activated when the RNA reaches its final destination. The evidence supporting this assumption in oligodendrocytes is circumstantial. If one process of an oligodendrocyte injected with GFP RNA is severed, GFP continues to be synthesized in the amputated process for several hours, indicating that the translation machinery is present and functional in the myelin compartment of oligodendrocytes (Barbarese et al. 1999).

One clue to understanding the mechanism for translation regulation during RNA trafficking comes from the observation that theA2RE functions as an enhancer of cap-dependent translation both in vivo and in vitro (Kwon et al. 1999; Fig. 5). The enhancer function is position- and copy number-independent and hnRNP A2-dependent. This means that, in addition to mediating nuclear export, granule assembly and transport along microtubules, A2RE/hnRNP A2 determinants also activate translation. Interaction of A2RE/hnRNP A2 determinants with the export machinery during nuclear export, with granule components during granule assembly and with the transport machinery during transport on microtubules may preclude their interaction with the translation machinery during early steps in the trafficking pathway. Once the RNA granule has reached the myelin compartment and dissociated from the transport machinery, A2RE/hnRNP A2 determinants may be free to interact with the translation machinery to activate translation (Fig. 6).

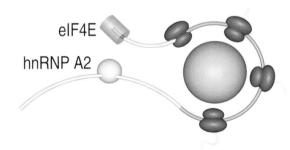

Fig. 6. Translation activation. HnRNP A2 associated with the A2RE in the 3'UTR of the RNA stimulates cap-dependent translation initiation at the 5' end of the RNA where eIF4E binds

11 Summary

A2RE and hnRNP A2 have been identified as important *cis/trans* determinants for MBP RNA trafficking in oligodendrocytes. Since A2RE-like sequences are found in several different transported RNAs, and since hnRNP A2 is expressed in most cell types, this may represent a general RNA trafficking pathway shared by a variety of different RNAs in different cell types. In oligodendrocytes, A2RE/hnRNP A2 determinants are involved in at least four steps in the RNA trafficking pathway: (1) export from the nucleus to the cytoplasm, (2) granule assembly in the perikaryon, (3) transport along microtubules in the processes, and (4) translation activation in the myelin compartment. The components of the cellular machinery mediating each of these steps are known. How A2RE/hnRNP A2 determinants interact with these components to mediate RNA trafficking is being investigated by a combination of: biochemistry to analyze molecular interactions in vitro, imaging to visualize molecular interactions in living cells, and computational modeling to simulate molecular interactions in the Virtual Cell.

References

Ainger K, Avossa D, Morgan F, Hill SJ, Barry C, Barbarese E, Carson JH (1993) Transport and localization of exogenous myelin basic protein mRNA microinjected into oligodendrocytes. J Cell Biol 123:431–441

Ainger K, Avossa D, Diana AS, Barry C, Barbarese E, Carson JH (1997) Transport and localization elements in myelin basic protein mRNA. J Cell Biol 138:1077–1087

Barbarese E, Koppel DE, Deutscher MP, Smith CL, Ainger K, Morgan F, Carson JH (1995) Protein translation components are colocalized in granules in oligodendrocytes. J Cell Sci 108:2781–2790

Barbarese E, Brumwell C, Kwon S, Cui H, Carson JH (1999) RNA on the road to myelin. J Neurocytol 28:263–270

Brumwell C, Antolik C, Carson JH, Barbarese E (2000a) Identification of a microtubule transport domain in hnRNP A2 (submitted)

Brumwell C, Carson JH, Barbarese E (2000b) Nuclear export of hnRNP A2 is facilitated by binding to an RNA trafficking sequence (submitted)

Carson JH, Worboys K, Ainger K, Barbarese E (1997) Translocation of myelin basic protein mRNA in oligodendrocytes requires microtubules and kinesin. Cell Motil Cytoskelet 38:318–328

Carson JH, Kwon S, Barbarese E (1998) RNA trafficking in myelinating cells. Curr Opin Neurobiol 8:607–612

Cui H, Xu H, Barbarese E, Carson JH (2000a) RNA sorting and assembly into granules is mediated by RNA trafficking determinants (submitted)

Cui H, Xu H, Barbarese E, Carson JH (2000b) Opposing motors regulate RNA trafficking (submitted)

Daneholt B (1997) A look at messenger RNP moving through the nuclear pore. Cell 88:585–588

Daneholt B (1999) Pre-mRNP particles: from gene to nuclear pore. Curr Biol 9:R412–R415

Dreyfuss G, Matunis MJ, Pinol-Roma S, Burd CG (1993) hnRNP proteins and the biogenesis of mRNA. Annu Rev Biochem 62:289–321

Gould RM, Freund CM, Barbarese E (1999) Myelin-associated oligodendrocytic basic protein mRNAs reside at different subcellular locations. J Neurochem 73:1913–1924

Hamilton BJ, Nichols RC, Tsukamoto H, Boado RJ, Partridge WM, Rigby WFC (1999) hnRNP A2 and hnRNP L bind the 3'UTR of glucose transporter 1 mRNA and exist as a complex in vivo. Biochem Biophys Res Commun 261:646–651

Hoek KS, Kidd GJ, Carson JH, Smith R (1998) hnRNP A2 selectively binds the cytoplasmic transport sequence of myelin basic protein mRNA. Biochemistry 37:7021–7029

Holz A, Schaeren-Wiemers N, Schaefer C, Pott U, Colello RJ, Schwab ME (1996) Molecular and developmental characterization of novel cDNAs of the myelin-associated/oligodendrocytic basic protein. J Neurosci 16:467–477

Izaurralde E, Jarmolowski A, Beisel C, Mattaj IW, Dreyfuss G, Fischer U (1997) A role for the M9 transport signal of hnRNP A1 in mRNA nuclear export. J Cell Biol 137:27–35

Kamma H, Horiguchi H, Wan L, Matsui M, Fujiwara M, Fujimoto M, Yazawa T, Dreyfuss G (1999) Molecular characterization of the hnRNP A2/B1 proteins: tissue-specific expression and novel isoforms. Exp Cell Res 246:399–411

Knowles RB, Sabry JH, Martone ME, Deerinck TJ, Ellisman MH, Bassell GJ, Kosik KS (1996) Translocation of RNA granules in living neurons. J Neurosci 16:7812–7820

Kwon S, Barbarese E, Carson JH (1999) The *cis*-acting RNA trafficking signal from myelin basic protein mRNA and its cognate *trans*-acting ligand hnRNP A2 function to enhance cap-dependent translation. J Cell Biol 147:247–256

Lall S, Francis-Lang H, Flament A, Norvell A, Schupbach T, Ish-Horowicz D (1999) Squid hnRNP protein promotes apical cytoplasmic transport and localization of *Drosophila* pair-rule transcripts. Cell 98:171–180

Lunn KF, Baas PW, Duncan ID (1997) Microtubule organization and stability in the oligodendrocyte. J Neurosci 17:4921–4932

Mayeda A, Munroe SH, Caceres JF, Krainer AR (1994) Function of conserved domains of hnRNP A1 and other hnRNP A/B proteins. EMBP J 13:5483–5495

Mouland AJ, Xu H, Cui, H, Krueger W, Munro TP, Prasol M, Mercier J, Rekosh D, Smith R, Barbarese E, Cohen EA, Carson JH (2000) Identification of RNA trafficking signals in the human immunodeficiency virus type 1 genome. Mol Cell Biol (in press)

Munro TP, Magee RJ, Kidd GJ, Carson JH, Barbarese E, Smith LM, Smith R (1999) Mutational analysis of a heterogeneous ribonucleoprotein A2 response element for RNA trafficking. J Biol Chem 274:34389–34395

Omoto C, Johnson KA (1986) Activation of the dynein adenosinetriphosphatase by microtubules. Biochemistry 25:419–427

Schaff J, Fink CC, Slepchenko B, Carson JH, Loew LM (1997) A general computational framework for modeling cellular structure and function. Biophys J 73:1135–1146

Siomi MC, Eder PS, Kataoka N, Wan L, Liu Q, Dreyfuss G (1997) Transportin-mediated nuclear import of heterogeneous nuclear RNP proteins. J Cell Biol 138:1181–1192

Visa N, Alzhanova-Ericsson AT, Sun X, Kiseleva E, Bjorkroth B, Wurtz T, Daneholt B (1996) A pre-mRNA-binding protein accompanies the RNA from the gene through the nuclear pores and into polysomes. Cell 84:253–264

Extrasomatic Targeting of MAP2, Vasopressin and Oxytocin mRNAs in Mammalian Neurons

Stefan Kindler, Evita Mohr, Monika Rehbein, and Dietmar Richter[1]

1 Introduction

In eukaryotic cells, individual cellular subregions are functionally specialized. This is reflected by an uneven distribution of particular organelles and molecular components. Neurons possess an intrinsic polarity by which intercellular input and output sites are separated to dendrites and axons, respectively. To maintain this structural and functional polarity it is crucial for a neuron to target individual proteins to their correct subcellular destination. Extrasomatic mRNA trafficking and translation is thought to contribute to subcellular protein sorting in neurons. Different isoforms of the microtubule-associated protein 2 (MAP2) and the respective mRNAs are both found in somata and dendrites of mammalian central nervous system neurons. MAP2 is known to influence microtubule stability and rigidity in vitro and in vivo. In neurons, MAP2 may be involved in regulating the morphology and function of dendritic spines and shafts. Extrasomatic transcript localization and local protein synthesis may provide a significant contribution to these regulatory processes. In magnocellular neurons of the hypothalamus, mRNAs encoding the vasopressin- and oxytocin-precursor protein are sorted to axons and dendrites. Both neuropeptides serve a dual function. After their secretion into the systemic circulation they act as hormones on different peripheral tissues. Moreover, they are released from dendrites into the central nervous system where they probably function as neurotransmitters/neuromodulators in events related to neuronal plasticity. In this chapter, we will focus on the recent identification of *cis*-acting extrasomatic targeting elements in MAP2, vasopressin and oxytocin mRNAs.

2 Microtubule-Associated Protein 2

2.1 Dendritic Localization of MAP2 mRNAs

In 1982, Steward and Levy showed that in the dentate gyrus polyribosomes are present in dendritic shafts of granule cells primarily at the base of dendritic

---

[1] Institute for Cell Biochemistry and Clinical Neurobiology, University of Hamburg, UKE, Martinistraße 52, 20246 Hamburg, Germany

spines. This observation indicated that in neurons mRNAs are trafficked to and translated in dendrites. This raised the question as to whether all neuronal mRNA classes are translocated from the soma into dendritic processes with a certain frequency or whether only selected transcripts are specifically targeted to dendrites. Subsequent studies seem to support the latter scenario (reviewed by Kindler et al. 1997; Steward 1997; Kuhl and Skehel 1998; Mohr 1999; Schuman 1999; Tiedge et al. 1999). Based on in situ hybridization data, most mRNAs in the vertebrate central nervous system appear to be restricted to neuronal somata, whereas a selected group of transcripts was also detected in dendrites. Garner et al. (1988) were the first to identify a specific mRNA in dendrites. Using in situ hybridization studies, they demonstrated that in the hippocampus and cortex of rat brain, MAP2a/b transcripts are not only present in neuronal somata, but are also detected in dendritic processes located in molecular layers (Garner et al. 1988; Tucker et al. 1989). Subsequently, it was shown that this subcellular mRNA distribution pattern of MAP2 mRNAs is preserved in cultured primary neurons derived from rat superior cervical ganglia (SCG) and hippocampus (Bruckenstein et al. 1990; Kleiman et al. 1990). Within dendrites, high levels of MAP2 transcripts in the proximal segments seem to decrease towards the tips of the cell processes (Garner et al. 1988; Tucker et al. 1989). Thus, the subcellular localization of MAP2 transcripts differs from the more even dendritic distribution pattern of other mRNAs, including those that encode the alpha subunit of the Ca^{2+}/calmodulin-dependent protein kinase II (α-CaMKII; Burgin et al. 1990), arg3.1/arc (Link et al. 1995; Lyford et al. 1995) and dendrin (Herb et al. 1997).

2.2 Distribution and Function of MAP2 in Neurons

Microtubules (MTs) are hollow cylindrical polymers that are built from α- and β-tubulin heterodimers (Mandelkow and Mandelkow 1995). They represent one of the major filament systems of the neuronal cytoskeleton. The core cylinder interacts with a group of microtubule-associated proteins (MAPs) that seem to stabilize and promote the assembly of MTs. MAP2 is the most abundant MAP in the mammalian brain. In sodium dodecylsulfate (SDS) polyacrylamide gels, it can be resolved into three distinct bands that migrate at 280, 270 and 70 kDa (Shafit-Zagardo and Kalcheva 1998). The corresponding isoforms that are named MAP2a, MAP2b and MAP2c, respectively, are translated from alternatively spliced mRNAs. Although MAP2 has initially been described as a neuron-specific protein, more recent reports suggest that minor amounts of individual isoforms are also present in glial cells (Shafit-Zagardo and Kalcheva 1998). In the rat brain, MAP2a, b and c exhibit distinct developmental expression patterns. Whereas MAP2b concentrations seem to be invariant throughout the entire lifetime, MAP2a appears first about 2 weeks after birth and remains detectable thereafter. In contrast, MAP2c seems to be present throughout embryonic development and almost totally disappears around postnatal day 14 (P14) (Johnson and Jope 1992).

All MAP2 isoforms possess a carboxyl-terminal MT-binding domain that contains either three or four imperfect amino acid repeats. In vitro MAP2 binding results in an increased stability and rigidity of MTs (Shafit-Zagardo and Kalcheva 1998; van Rossum and Hanisch 1999). Furthermore, the phosphorylation state of MAP2 modulates its regulatory role in microtubule dynamics (Brugg and Matus 1991; Illenberger et al. 1996; Itoh et al. 1997). In neurons, Ca^{2+} entry through N-methyl-D-aspartate (NMDA) type glutamate receptors can mediate a partial dephosphorylation of MAP2 via the activation of the phosphatase calcineurin (Quinlan and Halpain 1996). Thus, NMDA-receptor activity may influence MT dynamics.

The length of the amino-terminal MAP2 region that is not used for MT binding varies between isoforms. It seems to point away from the MT surface and is thus referred to as the projection domain. It is thought to serve at least two different functions. First, it seems to regulate the spacing between MTs. In vitro studies have shown that the short projection domain in MAP2c allows tighter packing of MTs than the longer amino terminus of MAP2b (Chen et al. 1992). Secondly, the projection domain interacts with other molecular components of the cytoplasm. The regulatory RII subunit of the cAMP-dependent protein kinase (PKA) has been shown to bind an amino-terminal sequence element in the projection domain of all MAP2 isoforms (Obar et al. 1989; Rubino et al. 1989). Although the corresponding binding domains on MAP2 have not been clearly described, actin, neurofilaments, and mitochondria may also interact with MAP2 (Wiche 1989).

In accordance with the subcellular distribution of its mRNA, MAP2 exhibits a somatodendritic localization pattern in neurons. However, the dendritic localization of MAP2 mRNAs does not seem to be a prerequisite for the correct subcellular distribution of the corresponding proteins. Although a recombinant mRNA encoding MAP2c was restricted to neuronal cell bodies in transgenic mice, the corresponding protein exhibited a somatodendritic distribution indistinguishable from the localization pattern of endogenous MAP2 isoforms (Marsden et al. 1996). Based on the specific accumulation of MAP2 in dendritic processes, it has been speculated that it plays an important role in organizing the cytoskeleton in dendrites. In addition, the projecting arm of MAP2 may act as a scaffold for signaling elements on the MT surface. MAP2 may also be involved in regulating the morphology of dendritic spines (van Rossum and Hanisch 1999). A number of data suggest that dendritic spines, especially in the developing brain, contain a highly labile population of microtubules (Westrum et al. 1980; Chicurel and Harris 1992). Electron microscopic studies indicate that MAP2 is also present in dendritic spines (Fifkova and Morales 1989; Morales and Fifkova 1989). Nonetheless, the presence of MTs and MAPs in these synaptic structures remains a controversial issue (Bernhardt and Matus 1984; Kaech et al. 1997; van Rossum and Hanisch 1999). Taken together, the above data suggest that modulations of the subcellular concentration and phosphorylation status of MAP2 may be involved in the control of the morphology and function of dendritic spines and shafts. The dendritic localization of MAP2 mRNAs in conjunction with an extrasomatic translation may contribute to the regulation of MAP2 levels in individual subdomains of dendrites. This hypothesis is supported by a recent report indicating that synaptic activity can modulate the MAP2 concentration in the molecular layer of the dentate gyrus in rat brain (Steward and Halpain 1999). High-frequency stimulation of perforant path projections induced a two-stage change in the MAP2 immunostaining intensity in the molecular layer. A 5-min stimulation led to a decreased staining signal in the dendritic lamina harboring the activated synapses. In contrast, after stimulation for 1–2 h, MAP2 immunostaining increased in proximal and distal dendritic regions flanking the zone of activated synapses. An inhibition of protein synthesis diminished, but did not eliminate, the described alterations in immunostaining. This suggests that the observed activity-induced increase in dendritic MAP2 levels results in part from a local protein synthesis in cell processes. Currently, the molecular means that direct a spatially restricted extrasomatic MAP2 synthesis are enigmatic. It will be interesting to determine whether the cytoplasmic polyadenylation element binding protein (CPEB) which has been implicated in regulating the αCaMKII transcript translation at synaptic sites (Wu et al. 1998) is also involved the control of a dendritic MAP2 biosynthesis.

2.3 Characterization of a *cis*-Acting Dendritic Targeting Element in the MAP2 3' UTR

Despite a steadily increasing number of transcripts detected in dendritic processes, our current understanding of the molecular constituents that are involved in this cytoplasmic translocation process is very limited. In analogy to cytoplasmic RNA transport phenomena investigated in nonneuronal cell systems, two basic groups of molecular components seem to direct dendrite-specific transcript sorting. First, dendritically localized mRNAs are likely to contain so-called *cis*-acting elements. These are nucleotide stretches of sorted transcripts that serve as intrinsic signals for the subcellular translocation process. Secondly, proteins referred to as *trans*-acting factors seem to interact with *cis*-acting elements. Subsequently, the resulting ribonucleoprotein complex may functionally couple to a cellular transport device that is potentially based on cytoskeletal filaments. To date, only very few studies aimed at the identification of *cis*- and *trans*-acting components of the dendritic mRNA transport machinery have been published. Using a transgenic approach, the 3' untranslated region (3' UTR) of α-CaMKII mRNAs has been shown to impart extrasomatic translocation competence on normally nondendritic β-galactosidase transcripts (Mayford et al. 1996). Subsequent studies to further delineate sequence elements within the α-CaMKII 3' UTR that direct dendritic mRNA localization have not yet been performed.

The data currently available with respect to the identification of a *cis*-acting dendritic targeting element (DTE) in MAP2 mRNAs are in part contradictory. A first indication towards the position of such a signal element within MAP2 transcripts came from in situ hybridization studies performed with rat hippocampal sections (Papandrikopoulou et al. 1989). Using an oligonucleotide that was designed to specifically interact with MAP2c transcripts, endogenous MAP2c mRNAs in the hippocampus of young and adult rats were found to be restricted to cell bodies. In contrast, in situ hybridization experiments performed with a longer cDNA probe that only hybridizes with unique sequences in the coding region of MAP2a/b transcripts suggested a somatodendritic distribution pattern of the corresponding mRNAs (Papandrikopoulou et al. 1989). Thus, these findings suggested that *cis*-acting elements for dendritic mRNA trafficking are located in unique regions of MAP2a/b mRNAs that are not found in the shorter MAP2c transcripts. In subsequent studies, we have characterized those unique regions in MAP2a/b transcripts (Kindler et al. 1990, 1996; Kindler and Garner 1994; Chung et al. 1996). These investigations show that the major difference between MAP2a/b mRNAs and MAP2c transcripts is the absence of a segment comprising about 4 kb from the coding region of the shorter messages. In contrast to the variance in the coding region, all MAP2 transcripts seem to contain identical sequences within their 3' UTRs. This conclusion is also consistent with the finding that in the human MAP2 gene, a single exon comprises almost the entire sequence information of the 3' UTR (Kalcheva et al. 1995). In the human brain, alternative splicing leads to the expression of at least three distinct transcripts that partially differ in their 5' UTR sequences (Kalcheva et al. 1995). It is not yet clear whether those distinct 5' sequences are specifically associated with long MAP2a/b and/or short MAP2c transcripts. Taken together, the data described above suggest that a *cis*-acting DTE resides in the 4-kb section of the coding region unique to the dendritically localized long MAP2a/b mRNAs (Kindler et al. 1996).

In the in situ hybridization study performed by Papandrikopoulou et al. (1989), the regional distribution of endogenous MAP2c mRNAs was analyzed with a radioactively labeled oligonucleotide specific for short MAP2c transcripts, whereas a labeled cDNA probe comprising about 1.5 kb was used to identify the subcellular localization of MAP2a/b mRNAs. Thus, one might speculate that the seemingly restricted somatic distribution pattern of MAP2c mRNAs may in part be due to a weak oligonucleotide-based hybridization signal in comparison to a stronger signal obtained with the long cDNA probe. To help resolve this issue, we have analyzed the spatial distribution of both MAP2c and MAP2a/b transcripts in rat hippocampus by performing an in situ hybridization analysis with two radioactively labeled oligonucleotides that either specifically recognize only ~5.6 kb MAP2c mRNAs or ~9.6 kb MAP2a/b transcripts. Using a Northern blot assay with total RNA from the brain of a 6-day-old rat, the assumed specificity of both oligonucleotides was confirmed (Fig. 1B). When the same oligonucleotides were used to perform in situ hybridizations with sagittal tissue sections from rat hippocampus, MAP2a/b mRNAs were found to exhibit a somatodendritic distribution pattern similar to the subcellular localization that has been described before (Fig. 1C, d; Garner et al. 1988; Papandrikopoulou et al. 1989; Tucker et al. 1989). However, in con-

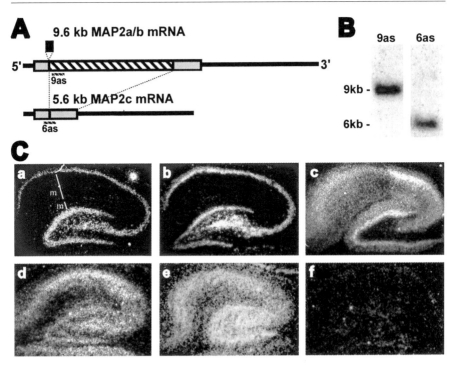

Fig. 1. **A** Schematic representation of the structural organization of ~9.6 kb MAP2a/b and ~5.6 kb MAP2c mRNAs. *Boxes* and *solid black lines* indicate coding and untranslated regions, respectively. Protein coding regions specific for MAP2 a/b (*hatched box*) and MAP2a (*black box*) mRNAs are shown. Regions hybridizing with oligonucleotide probes specific for either high (9as) or low molecular weight transcripts (6as) are indicated. **B** Northern blot with total RNA from the brain of a 5-day-old rat (P5) showing the specificity of both oligonucleotides. **C** In situ hybridization experiments with sagittal hippocampal sections from P7 (**b, d, e**, and **f**) and P10 (**a** and **c**) rat brain and oligonucleotide probes specific for glycerine aldehyde 3-phosphate dehydrogenase (GAPDH; **a**), α-tubulin (**b**), α-CaMKII (**c**), MAP2a/b (**d**), and MAP2c (**e**) mRNAs. **f** Control experiment performed with a sense probe complementary to the oligonucleotide 6as. Whereas α-tubulin and GAPDH transcripts are restricted to cell body layers of the hippocampus and dentate gyrus, α-CaMKII, MAP2a/b, and MAP2c mRNAs are also found in molecular (*m*) layers

trast to the data reported by Papandrikopoulou et al. (1989), in situ hybridization studies performed with a MAP2c-specific oligonucleotide probe resulted in a signal located over molecular as well as cell body layers, indicating a somatodendritic compartmentalization of these transcripts (Fig. 1C, e). Thus, our data hint towards the localization of a *cis*-acting DTE in common regions shared between all MAP2 transcripts, excluding the unique 4 kb segment in the coding region of MAP2a/b mRNAs.

To resolve this issue, we have recently established two different assay systems to identify *cis*-acting dendritic trafficking elements of MAP2 mRNAs in a functional manner (Blichenberg et al. 1999). It was assumed that as part of a chimeric transcript, a *cis*-acting DTE should be capable of imparting extrasomatic localization competence on exogenous, somatically restricted reporter mRNAs. In a series of experiments, a set of eukaryotic expression vectors was introduced into primary hippocampal and sympathetic SCG neurons by transfection and microinjection, respectively. In cell culture, both neuronal cell types are known to extend axonal and dendritic processes (Blichenberg et al. 1999) as well as localize endogenous MAP2 transcripts to dendrites (Bruckenstein et al. 1990; Kleiman et al. 1990). The basic expression plasmid led to the transcription of an mRNA containing sequences that encode the enhanced green fluorescent protein (EGFP). EGFP mRNAs are not endogenously expressed in mammalian neurons. Moreover, previous experiments have shown that somatically injected EGFP transcripts remain restricted to the cell body of sympathetic SCG neurons (Muslimov et al. 1997). Additional vectors were constructed that led to the synthesis of hybrid mRNAs containing EGFP-encoding sequences and various regions of MAP2 transcripts. In another vector, the cDNA representing the entire α-tubulin transcript was inserted downstream of the EGFP sequence into the basic expression vector. Endogenous as well as recombinant α-tubulin mRNAs transcribed from eukaryotic expression vectors were previously shown to remain restricted to neuronal somata (Bruckenstein et al. 1990; Prakash et al. 1997). The subcellular localization of chimeric mRNAs was monitored by nonradioactive in situ hybridization using a digoxygenin-labeled riboprobe that specifically interacts with the EGFP sequences in the recombinant transcripts. Thereby, all hybrid mRNAs could be detected with an identical sensitivity level and distinguished from endogenous MAP2 transcripts. In general, data obtained with both hippocampal and sympathetic cells were very similar.

Theoretically, *cis*-acting signals of targeted mRNAs can act in at least two distinct ways. Data obtained in various nonneuronal cell systems indicate that *cis*-acting trafficking elements on localized transcripts must interact with *trans*-acting RNA-binding proteins to mediate cytoplasmic mRNA sorting. Alternatively, *cis*-elements may encode protein targeting domains. Thus, a translational complex may be localized by virtue of a sorting signal encoded on the nascent polypeptide leading to a passive co-transport of the corresponding mRNA (Marsden et al. 1996). The first set of eukaryotic expression vectors that were introduced into primary neurons were used to investigate whether dendritic MAP2 transcript trafficking depends on sorting signals contained in mRNAs or proteins. In the first chimeric transcripts, the 5′ UTR and the entire coding region of MAP2b mRNAs were located downstream of the EGFP sequence such that MAP2 sequences were not translated (vector pNEcu). These hybrid transcripts remained in the cell soma. In the second reporter mRNA, the MAP2a/b coding region was situated upstream and in frame with the EGFP sequence (vector pNEc). These transcripts that led to the synthesis of MAP2/EGFP fusion proteins were also retained in the cell soma. In contrast, hybrid mRNAs carrying the entire MAP2 3′ UTR downstream of the EGFP coding region were localized into dendrites in the majority of microinjected sympathetic and transfected hippocampal neurons (vector pNEu). Control transcripts consisting of the EGFP coding region fused to the entire α-tubulin mRNA were restricted to cell somata and proximal-most dendritic segments of both types of primary neurons. To quantify the described observations, cells that were labeled during the in situ hybridization were grouped into two classes. First, those exhibiting a 'somatic' mRNA distribution pattern for which the hybridization signal was restricted to somata and proximal-most sections of dendrites, and secondly, cells in which chimeric transcripts were 'dendriti- cally localized' to parts of at least one neuronal process that were further away from the soma than one cell body diameter as measured from the base of the dendrite. Based on this classification, hybrid mRNAs containing the MAP2 3′ UTR were dendritically localized in about 60% of the investigated sympathetic and hippocampal neurons. However, chimeric transcripts containing α-tubulin RNA sequences or MAP2 5′ untranslated and coding regions displayed almost exclusively a somatic distribution pattern (>96%). Chimeric mRNAs were never observed in axons. In conclusion, the described observations that are summarized in Fig. 2 show that in sympathetic and hippocampal neurons, MAP2 mRNAs are selectively and specifically sorted into dendrites. Dendritic transcript trafficking is directed by a cis-acting targeting signal situated in the MAP2 3′ UTR and is not indirectly mediated by an inherent subcellular sorting domain of the nascent polypeptide chain.

The MAP2 3′ UTR comprises about 3.74 kb (Kindler et al. 1996). It seems likely that this portion of the mRNA does not only play a role in subcellular transcript localization, but may also direct additional cellular functions such as RNA stability and translation. To further delineate the cis-acting DTE in the 3′ UTR, a number of hybrid transcripts containing subfragments of this region were expressed in primary neurons. The results obtained with hippocampal and sympathetic neurons are summarized in Fig. 2B. In brief, a 640-nucleotide region comprising nucleotides 2432–3071 of the MAP2 3′ UTR possesses all the essential cis-acting signals to sufficiently mediate dendritic mRNA trafficking. Additional 5′ and 3′ deletions ($pNEu_{2632-3071}$ and $pNEu_{2432-2807}$), interfered dramatically with the extrasomatic targeting capacity of the characterized cis-element. This indicates that the entire 640 nucleotide DTE is required for dendritic transcript localization. The primary significance of the DTE for extrasomatic mRNA targeting was subsequently corroborated by its deletion from the intact 3′ UTR ($pNEu\Delta_{2436-3071}$). Distinct from the mainly dendritic distribution pattern of chimeric transcripts comprising the complete 3′ UTR (pNEu), mRNAs transcribed from $pNEu\Delta_{2436-3071}$ were almost exclusively refined to cell somata. Thus, a 640-nucleotide DTE residing in the MAP2 3′ UTR emerges as a cis-acting element that is both sufficient and essential in efficient trafficking of MAP2 mRNAs into dendrites of primary neurons. In spite of methodological variations, essentially identical results were obtained in hippocampal and sympathetic cell systems, indicating that a central component of the dendritic mRNA targeting machinery functions in both cell types.

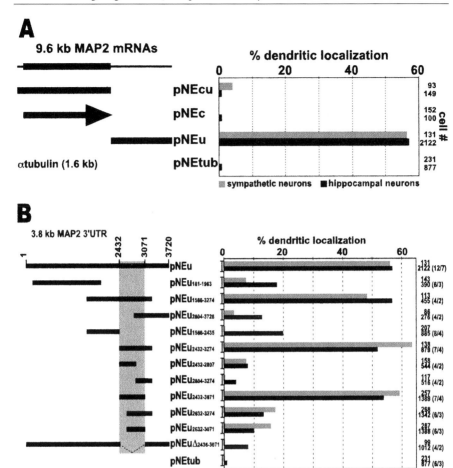

Fig. 2. A Subcellular distribution of chimeric transcripts in sympathetic and hippocampal neurons. In the *left panel*, the molecular structure of 9.6 kb MAP2 transcripts is schematically indicated with *black bars* and *lines* representing coding and noncoding regions, respectively. *Below* MAP2 sequences that are present in recombinant mRNAs derived from the corresponding vectors are indicated as *heavy black bars*. From pNEc a MAP2/EGFP fusion protein is synthesized (indicated by a *black arrow*), whereas the MAP2 coding region in pNEcu transcripts is not translated into a protein. The *right panel* lists the relative amount of cells showing dendritic localization of chimeric transcripts in sympathetic and hippocampal neurons and the total number of evaluated cells per construct. In both types of transgenic primary neurons, only recombinant mRNAs containing the MAP2 3′ UTR are efficiently localized to dendrites. B Identification of a 640-nucleotide *cis*-acting dendritic targeting element (DTE) in the 3′ UTR of MAP2 transcripts. In the *left panel*, regions of the MAP2 3′ UTR included in chimeric mRNAs transcribed from the corresponding vectors are shown as *heavy black bars*. In the *right panel*, the percentage of primary neurons exhibiting dendritic mRNA localization patterns and the total number of analyzed cells are shown. In parentheses, the total number of investigated coverslips/independent transfections of hippocampal neurons are shown. Only chimeric transcripts containing a 640-nucleotide-long sequence highlighted in *gray* are detected with high frequency in dendrites of sympathetic and hippocampal neurons. Deletion of this targeting element eliminates dendritic localization. (Reproduced by courtesy of Blichenberg et al. 1999)

A number of observations in nonneuronal cell systems suggest that a specific secondary structure of the RNA serves as a cellular sorting signal (Macdonald 1990; Chartrand et al. 1999). Thus, we used the mfold program (Zuker 1989) to predict an optimal secondary structure for the DTE on the basis of free energy minimization. The DTE is predicted to form a complex stem and loop pattern shown in Fig. 3A. Interestingly, the secondary structure of a core region of the DTE (area highlighted in gray in Fig. 3B) comprising about 520 nucleotides (~2543–3066) also seems to be conserved in the context of the entire 3′ UTR sequence. 5′ or 3′ deletions into the DTE core region that have been shown to disrupt its dendritic localization capacity (pNEu$_{2632-3071}$ and pNEu$_{2432-2807}$ transcripts, respectively) are likely to disrupt the central stalk of this structure (Fig. 3A). Future experiments will be necessary to evaluate the significance of individual structural elements within the DTE for dendritic RNA localization.

3 Axonal and Dendritic Targeting of VP and OT mRNAs

The hypothalamus is a brain region involved in neuroendocrine functions. With regard to subcellular mRNA sorting, this brain structure deserves some interest, since it is composed of neurons harboring protein-encoding transcripts in their axons and dendrites. Data summarized here will focus on subcellular targeting of vasopressin (VP) and oxytocin (OT) mRNAs. In vivo, the genes encoding these peptide hormone precursors are expressed in different populations of magnocellular neurons of the hypothalamic-neurohypophyseal system. Apparently, both neuropeptides have a dual function. They are secreted from the nerve terminals in the posterior pituitary into the systemic circulation to act as hormones on their different target tissues in the periphery. In addition, VP and OT are also released from dendrites into the central nervous system where they presumably play a role as neurotransmitters/neuromodulators in processes related to neuronal plasticity as a consequence of external stimulation (reviewed in Morris et al. 1993).

3.1 Axonal mRNA Compartmentalization

Characterization of the axonally localized VP and OT transcripts has revealed several interesting features:

1. The mRNAs exhibit shorter poly(A) tracts than their counterparts in the cell bodies. The poly(A) tail length of both VP and OT mRNAs in the cell bodies may vary considerably in response to external physiological stimulation, for instance following osmotic challenge. In contrast, no such variation has been observed for these RNAs in the distal parts of axons in the posterior pituitary. The mRNAs are identical in size in osmotically stressed and control animals.

A

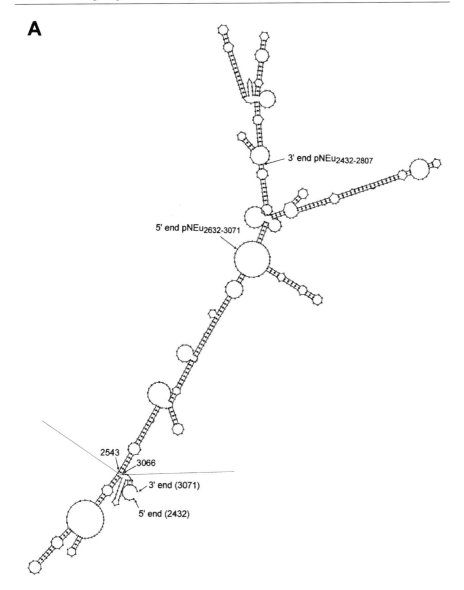

3' end pNEu$_{2432-2807}$

5' end pNEu$_{2632-3071}$

2543

3066

3' end (3071)

5' end (2432)

Fig. 3A,B. Predicted optimal secondary structure of the 640-nucleotide DTE (**A**) and the MAP2 RNA sequences present in transcripts from the vector pNEu including the entire MAP2 3' UTR (**B**). Structures were predicted with the program MFOLD. **A** The localization of the 5' and 3' ends of pNEu$_{2632-3071}$ and pNEu$_{2432-2807}$ transcripts are marked. *Numbers* indicate nucleotide positions according to the entire 3' UTR sequence. The structure of RNA parts lying above the *gray line* in **A** are conserved in the structure predicted for the entire 3' UTR and are highlighted in *gray* (**B**). **B** The position of the 5' and 3' ends of pNEu transcripts and the DTE are indicated. Images of predicted secondary RNA structures were kindly provided by Arne Blichenberg at the Institute for Cell Biochemistry and Clinical Neurobiology, Hamburg, Germany (Fig. 3B see page 94)

B

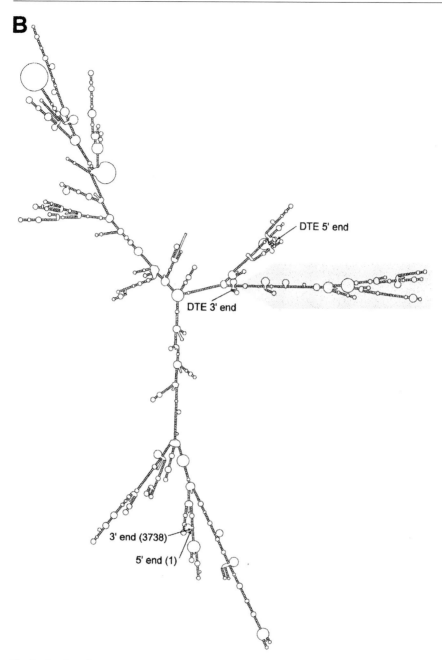

Fig. 3. *Continued.*

2. Osmotic stress leads to a differential accumulation of VP and OT transcripts in the axonal compartment. When rats are subjected to salt loading, a substantial increase in the amount of axonal VP mRNA (approximately 17-fold) is observed exceeding the two- to threefold increase which takes place in the cell bodies. A much lower accumulation rate is observed for the axonal OT-encoding transcripts (threefold) which resembles the increase of this RNA species in the cell bodies (reviewed in Mohr and Richter 1995).

3. Trembleau et al. (1996) have shown that, within a given axon in the posterior pituitary, the VP-containing neurosecretory vesicles and VP mRNA are not co-localized. Rather, axonal varicosities immunoreactive for VP are devoid of VP mRNA and vice versa. These data suggest that VP and OT mRNAs are not passively spilling over to the axonal compartment as a consequence of the high secretory activity of peptidergic neurons.

4. The mRNAs are transported to the axonal compartment subsequent to translation. Evidence stems from investigating the subcellular distribution of VP-encoding mRNA in the homozygous and heterozygous Brattleboro (BB) rat. Due to a frame-shift mutation in the VP gene, translation of the mutant hormone precursor protein is extremely inefficient. VP mRNA is undetectable in axons of affected homozygous animals. In heterozygous animals in which both the wild type and the mutant alleles are transcribed, the wild-type mRNA is sorted to axons whereas the mutant transcripts remain confined to the cell somata (Mohr et al. 1995). The translational termination of the mutant transcript may be impaired because it lacks any in-frame stop-codons. This in turn might eventually lead to a local degradation at the rough endoplasmic reticulum (RER) and thus to a failure of targeting to the axon.

5. Evidence for the association of axonal VP and OT mRNAs with ribosomes or polysomes is lacking. Hence, local translation into the precursor proteins is unlikely to occur (Mohr et al. 1995).

Hypothalamic magnocellular neurons are not unique in harboring axonal mRNAs. Other examples include neurons of the olfactory bulb. Various mRNA species, among them transcripts encoding the olfactory marker protein and several odorant receptors, are located in distal axonal segments while equally abundant RNAs are restricted to the perikarya (Ressler et al. 1994; Vassar et al. 1994; Wensley et al. 1995). Axons are generally believed to lack protein synthesis capacity (Lasek and Brady 1981). An exception to this dogma is the initial axonal segment which contains components of the translational machinery (reviewed in Steward et al. 1995). Hence, transport of mRNA encoding the tau protein to the initial segment in cultured rat cortical neurons (Litman et al. 1993) may indeed allow local on-site synthesis of the corresponding protein. Even though the available data suggest that compartmentalization of mRNAs to the axonal domain of certain nerve cell types is a specific process, the physiological meaning of the majority of mRNAs in this cellular compartment has still to be defined.

3.2 Dendritic mRNA Compartmentalization

A decade ago VP- and OT-encoding transcripts were observed in neurites of hypothalamic magnocellular neurons with the morphological appearance of dendrites (Bloch et al. 1990). Since then, these data have been confirmed by more extensive characterization of dendritic VP and OT mRNAs. When compared with mRNAs residing in the axonal compartment, dendritic transcripts exhibit distinct characteristics: (1) poly(A) tail polymorphism has not been observed; dendritic RNAs and those residing in the cell somata appear to be identical in size, and (2) VP mRNA is targeted to the dendrites in homozygous BB rats, emphasizing that mRNA targeting to axons and dendrites, respectively, is likely to be brought about by distinct molecular mechanisms (Mohr et al. 1995). Furthermore, the data suggest dendritic RNA transport to take place prior to translation. Immunohistochemical studies at the ultrastructural level clearly demonstrate synthesis of the VP and OT precursors in small cisterns of RER in dendrites of hypothalamic magnocellular neurons (Morris et al. 1997). By employing electron microscopic in situ hybridization techniques, VP mRNA was detected in dendritic segments containing RER (Prakash et al. 1997), strongly suggesting that local on-site synthesis of at least the VP precursor may take place. A major question remains to be answered, namely, how may secretory or membrane-associated proteins be synthesized in dendrites? It is still a matter of debate whether or not dendrites harbor Golgi-like structures. Even though some Golgi marker proteins appear to be present, these molecules are located in only some dendrites of a given neuron and most often in parts very proximal to the cell body (Tiedge and Brosius 1996; Torre and Steward 1996).

3.3 Molecular Determinants of Subcellular mRNA Targeting

A heterologous system, namely primary cultured SCG neurons derived from embryonic rats, has proven useful to define the *cis*-acting signals required for VP and OT mRNA targeting to the neuronal processes (the peptide hormone genes are not endogenously expressed in these neurons). When eukaryotic expression vectors containing the cDNAs in sense orientation were introduced into these cells by nuclear microinjections, VP and OT transcripts were detectable throughout the cell somata as well as in dendrites (Prakash et al. 1997). These data imply that primary cultured SCG neurons are equipped with components necessary for dendritic sorting of mRNAs that are not endogenously expressed within these cells. Apparently, at least some of the molecular determinants of the mRNA localization machinery are not cell-specific. This view is consistent with observations made in non-neuronal cells. For instance, mRNAs encoding isoforms of actin are correctly sorted to different subcellular locations in various cell types. Transfection studies have revealed the correct intracellular localization of α-, β- and γ-actin mRNAs regardless of the cell type employed for transfection and whether or not a particular isoform was endogenously expressed in that cell type (Kislauskis et al. 1995). As mentioned above, VP and OT mRNAs are sorted to dendrites as well as to axons in hypothalamic magnocellular neurons. In microinjected SCG neurons, however, these transcripts are not detectable in the axonal compartment (Prakash et al. 1997). Only in rare cases could some cells be observed harboring VP transcripts in the very proximal axonal segment. Consequently, molecules necessary for sorting of distinct mRNA species to the axon may be restricted to a limited number of cell types such as hypothalamic magnocellular neurons and a few other nerve cells. Alternatively, since the axons of SCG neurons are very long and form a dense network, low levels of mRNA might escape detection.

Many investigations made recently, particularly in *Drosophila* and *Xenopus* oocytes and early embryos, have demonstrated the importance of defined and rather complex targeting elements in the RNA molecule to be localized which play a critical role in the formation of a functional transport unit (reviewed in Bashirullah et al. 1998). In many mRNAs, these *cis*-acting elements are located in the 3′ UTR and they are necessary and sufficient for the correct temporal and spatial subcellular compartmentalization of the respective transcript. Concerning the dendritic transport of VP and OT mRNAs in microinjected SCG neurons, however, sequences within the coding region in addition to the 3′ UTR are required for their efficient sorting to the dendrites (Prakash et al. 1997; Mohr 1999). The dendritic targeting capacity of various segments spanning the VP RNA has been analyzed by ligating parts of the VP cDNA individually to the 3′ end of α-tubulin cDNA such that they form part of the 3′ UTR. Unmodified α-tubulin transcripts are confined to the cell bodies regardless of whether they were derived by transcription of the endogenous genome or by that of a microinjected expression vector (Prakash et al. 1997). Numerous experiments of that kind, summarized in Fig. 4, have revealed that individual subsegments may indeed confer a moderate degree of dendritic targeting capacity to the recombinant RNA. The full extent of dendritic RNA transport was only restored when a longer fragment spanning part of the coding region as well as the 3′ UTR was ligated to the α-tubulin construct. Thus, the dendritic localizer elements within VP mRNA are obviously redundant and all of them must be combined in order to achieve an efficient mRNA targeting to dendrites.

4 Final Conclusions and Perspectives

In neurons, dendritically localized mRNAs may be associated with other cellular components. When neuronal cultures derived from embryonic mouse brain were used to biochemically separate polymerized microtubules from unassembled tubulin, the MAP2 mRNA concentration in the microtubule fraction was about 5 times higher than in the soluble fraction (Litman et al. 1994). Upon depolymerization of microtubules by colchicin treatment, MAP2 mRNAs were released from the filaments. These data suggest that in the neuronal cyto-

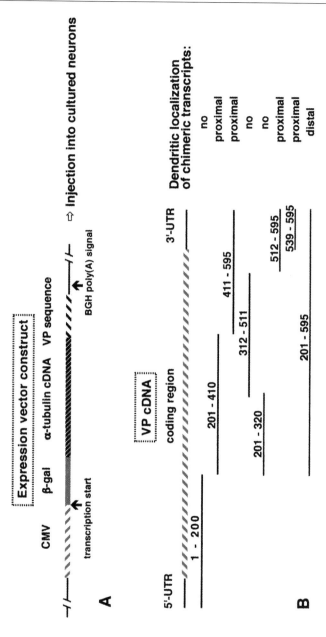

plasm, MAP2 mRNAs are bound to microtubules. In addition, in situ hybridization signals of dendritically localized recombinant mRNAs containing MAP2, VP and OT sequences appeared in patches that were scattered along the length of cell processes (Fig. 5). This may reflect a discontinuous nature of the extrasomatic mRNA targeting apparatus, with individual migrating transcript granules, or alternatively, with mRNAs preferentially docking to particular binding sites along dendritic shafts. Interestingly, after fluorescent labeling of endogenous transcripts in cultured hippocampal neurons (Knowles et al. 1996) as well as microinjection of labeled RNA into primary SCG neurons (Muslimov et al. 1997) and in cultured oligodendrocytes (Ainger et al. 1993), similar extrasomatic RNA granules were observed. The granules may contain macromolecules other than the mRNA to be transported, for instance components of the translational machinery such as ribosomes, elongation factors and aminoacyl-tRNA synthetases (Barbarese et al. 1995). It is conceivable that for the definition of microdomains, mRNA sorting within a given cell may be mediated by the formation of distinct macromolecular transport units. Their exact composition and mode of assembly remain to be characterized in more detail.

To date, only very few studies to functionally characterize *cis*-acting sorting signals of extrasomatic transcripts in neurons have been performed. The data summarized here show that at least some *cis*-acting signals involved in dendritic mRNA sorting are functional in distinct neuronal cell types. Whereas DTEs in MAP2 and α-CaMKII mRNAs appear to primarily reside in the 3′ UTR, dendritic localizer elements in VP transcripts seem to be spread over 3′ noncoding as well as coding regions. Moreover, several *cis*-acting signals located in different sections of the VP mRNA seem to act synergistically to

◄

Fig. 4A,B. Schematic representation of dendritic localizer elements within vasopressin (VP) mRNA. To define the sequences mediating dendritic transport of VP mRNA, eukaryotic expression vectors were designed and subsequently microinjected into cultured neurons. The expression vector is schematically shown in **A**. Expression of any inserted cDNA is driven by the cytomegalovirus (CMV) promoter. A short oligonucleotide derived from the bacterial β-galactosidase gene (β-gal) was inserted such that it forms part of the 5′-untranslated region (UTR) of vector-expressed mRNAs. It allows subsequent detection of vector-expressed RNAs with a universal β-gal-specific anti-sense oligonucleotide. Individual parts of the rat VP cDNA (nucleotide positions are indicated by *numbers*), schematically depicted in **B**, were ligated to the 3′ end of rat α-tubulin cDNA. The VP sequence forms part of the 3′ UTR of the chimeric mRNAs. The bovine growth hormone (BGH) poly(A) signal allows for the addition of a poly(A) tail to resulting transcripts. The subcellular distribution of vector-expressed RNAs was analyzed by non-radioactive in situ hybridization procedures. Dendritic localizer elements within VP RNA, as summarized on the *right* of part B, are redundant and they are located in the coding region as well as within the 3′ UTR. It is obvious that individual subfragments from the region spanning nucleotide positions 201–595 were only able to direct the chimeric mRNAs to the proximal parts of the dendrites. The full extent of dendritic targeting capacity to distal locations was mediated by a fragment spanning the complete nucleotide sequence from position 201–595, indicating a synergistic action of the various localizer elements. For experimental details see Prakash et al. (1997). (Reprinted from Mohr 1999, pp. 507–525, with permission from Elsevier Science)

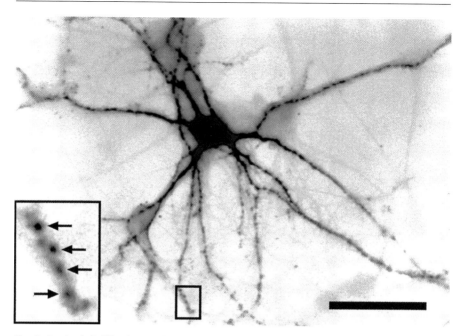

Fig. 5. In situ hybridization analysis of the subcellular distribution of chimeric mRNAs transcribed from pNEu$_{2432-3071}$ in hippocampal neurons. Bright field micrograph of a cell expressing recombinant transcripts containing the MAP2-DTE. Higher magnification image shown as inset depicts the granular nature of the color-reaction product of the non-radioactive in situ hybridization in dendrites (*arrows*). *Scale bar* 50 μm

promote dendritic RNA trafficking. In MAP2 mRNAs, a single element comprising 640 nucleotides appears to be both sufficient and essential for dendritic transcript trafficking. However, *cis*-elements situated in other parts of the mRNA could contain accessory targeting signals. Interestingly, extrasomatic sorting signals in MAP2 and vasopressin transcripts do not share any striking sequence similarity. Moreover, they are not remarkably similar to 3′ UTR sequences of α-CaMKII mRNAs. Although the presence of rather short conserved sequence elements that are involved in subcellular transcript sorting cannot be excluded, distinct transcripts may use slightly different molecular mechanisms to couple to a dendritic trafficking system. It is also important to emphasize that the sequence of *cis*-elements per se may not be the most significant characteristic for dendritic transcript sorting. Secondary and higher-order RNA structures may be key elements for extrasomatic trafficking, as has been shown for other cellular functions (Draper 1995; Conn and Draper 1998).

Despite recent advances in characterizing molecular components of the trafficking machinery, it remains a puzzle as to why transcripts encoding MAP2, VP and OT belong to a selected group of mRNAs that are extrasomatically localized in neurons. Although it is conceivable to assume that a spatially restricted synthesis of these proteins in dendrites may play a role in cellular events like synaptic plasticity, such a function has not been clearly demonstrated yet. Considering the continuous progress in the development of novel molecular and cell biological approaches, it seems likely that future experiments will shed light on the cell biological relevance of extrasomatic mRNA trafficking in neurons.

Acknowledgements. This research was supported by the Deutsche Forschungsgemeinschaft (Ri191–19–1, Ri192–21–1, FOG 296/2-1-5 and 296/2-1-4).

References

Ainger K, Avossa D, Morgan F, Hill SJ, Barry C, Barbarese E, Carson JH (1993) Transport and localization of exogenous myelin basic protein mRNA microinjected into oligodendrocytes. J Cell Biol 123:431–441

Bernhardt R, Matus A (1984) Light and electron microscopic studies of the distribution of microtubule-associated protein 2 in rat brain: a difference between dendritic and axonal cytoskeletons. J Comp Neurol 226:203–221

Barbarese E, Koppel DE, Deutscher MP, Smith CL, Ainger K, Morgan F, Carson JH (1995) Protein translation components are colocalized in granules in oligodendrocytes. J Cell Sci 108:2781–2790

Bashirullah A, Cooperstock RL, Lipshitz HD (1998) RNA localization in development. Annu Rev Biochem 67: 335–394

Blichenberg A, Schwanke B, Rehbein M, Garner C, Richter D, Kindler S (1999) Identification of a *cis*-acting dendritic targeting element in MAP2 mRNAs. J Neurosci 19:8818–8829

Bloch B, Guitteny AF, Normand E, Chouham S (1990) Presence of neuropeptide messenger RNAs in neuronal processes. Neurosci Lett 109:259–264

Bruckenstein DA, Lein PJ, Higgins D, Fremeau RT Jr (1990) Distinct spatial localization of specific mRNAs in cultured sympathetic neurons. Neuron 5:809–819

Brugg B, Matus A (1991) Phosphorylation determines the binding of microtubule-associated protein 2 (MAP2) to microtubules in living cells. J Cell Biol 114:735–743

Burgin KE, Waxham MN, Rickling S, Westgate SA, Mobley WC, Kelly PT (1990) In situ hybridization histochemistry of Ca^{2+}/calmodulin-dependent protein kinase in developing rat brain. J Neurosci 10:1788–1798

Chartrand P, Meng XH, Singer RH, Long RM (1999) Structural elements required for the localization of ASH1 mRNA and of a green fluorescent protein reporter particle in vivo. Curr Biol 9:333–336

Chen J, Kanai Y, Cowan NJ, Hirokawa N (1992) Projection domains of MAP2 and tau determine spacings between microtubules in dendrites and axons. Nature 360:674–677

Chicurel ME, Harris KM (1992) Three-dimensional analysis of the structure and composition of CA3 branched dendritic spines and their synaptic relationships with mossy fiber boutons in the rat hippocampus. J Comp Neurol 325:169–182

Chung WJ, Kindler S, Seidenbecher C, Garner CC (1996) MAP2a, an alternatively spliced variant of microtubule-associated protein 2. J Neurochem 66:1273–1281

Conn GL, Draper DE (1998) RNA structure. Curr Opin Struct Biol 8:278–285

Draper DE (1995) Protein-RNA recognition. Annu Rev Biochem 64:593–620

Fifkova E, Morales M (1989) Calcium-regulated contractile and cytoskeletal proteins in dendritic spines may control synaptic plasticity. Ann NY Acad Sci 568:131–137

Garner CC, Tucker RB, Matus A (1988) Selective localization of messenger RNA for cytoskeletal protein MAP2 in dendrites. Nature 336:674–677

Herb A, Wisden W, Catania M, Marechal D, Dresse A, Seeburg P (1997) Prominent dendritic localization in forebrain neurons of a novel mRNA and its product, dendrin. Mol Cell Neurosci 8:367–374

Illenberger S, Drewes G, Trinczek B, Biernat J, Meyer HE, Olmsted JB, Mandelkow EM, Mandelkow E (1996) Phosphorylation of microtubule-associated proteins MAP2 and MAP4 by the protein kinase p110mark. Phosphorylation sites and regulation of microtubule dynamics. J Biol Chem 271:10834–10843

Itoh TJ, Hisanaga S, Hosoi T, Kishimoto T, Hotani H (1997) Phosphorylation states of microtubule-associated protein 2 (MAP2) determine the regulatory role of MAP2 in microtubule dynamics. Biochemistry 36:12574–12582

Johnson GV, Jope RS (1992) The role of microtubule-associated protein 2 (MAP-2) in neuronal growth, plasticity, and degeneration. J Neurosci Res 33:505–512

Kaech S, Fischer M, Doll T, Matus A (1997) Isoform specificity in the relationship of actin to dendritic spines. J Neurosci 17:9565–9572

Kalcheva N, Albala J, O'Guin K, Rubino H, Garner C, Shafit-Zagardo B (1995) Genomic structure of human microtubule-associated protein 2 (MAP-2) and characterization of additional MAP-2 isoforms. Proc Natl Acad Sci USA 92:10894–10898

Kindler S, Garner CC (1994) Four repeat MAP2 isoforms in human and rat brain. Mol Brain Res 26:218–224

Kindler S, Schwanke B, Schulz B, Garner CC (1990) Complete cDNA sequences encoding rat high and low molecular weight MAP2. Nucleic Acids Res 18:2822

Kindler S, Müller R, Chung WJ, Garner CC (1996) Molecular characterization of dendritically localized transcripts encoding MAP2. Mol Brain Res 36:63–69

Kindler S, Mohr E, Richter D (1997) Quo vadis: extrasomatic targeting of neuronal mRNAs in mammals. Mol Cell Endocrinol 128:7–10

Kislauskis EH, Ross A, Latham VM, Zhu X, Bassell G, Taneja KL, Singer RH (1995) The mechanisms of RNA localization: its effect on cell polarity. In: Lipshitz HD (ed) Localised RNAs. Springer, Berlin Heidelberg New York, pp 185–195

Kleiman R, Banker G, Steward O (1990) Differential subcellular localization of particular mRNAs in hippocampal neurons in culture. Neuron 5:821–830

Knowles RB, Sabry JH, Martone ME, Deerinck TJ, Ellisman MH, Bassell GJ, Kosik KS (1996) Translocation of RNA granules in living neurons. J Neurosci 16:7812–7820

Kuhl D, Skehel P (1998) Dendritic localization of mRNAs. Curr Opin Neurobiol 8:600–606

Lasek RJ, Brady ST (1981) The axon: a prototype for studying expressional cytoplasm. Cold Spring Harbor Symp Quant Biol 46:113–124

Link W, Konietzko U, Kauselmann G, Krug M, Schwanke B, Frey U, Kuhl D (1995) Somatodendritic expression of an immediate early gene is regulated by synaptic activity. Proc Natl Acad Sci USA 92:5734–5738

Litman P, Barg J, Rindzoonski L, Ginzburg I (1993) Subcellular localization of tau mRNA in differentiating neuronal cell culture: implications for neuronal polarity. Neuron 10:627–638

Litman P, Barg J, Ginzburg I (1994) Microtubules are involved in the localization of tau mRNA in primary neuronal cell culture. Neuron 13:1463–1474

Lyford GL, Yamagata K, Kaufmann WE, Barnes CA, Sanders LK, Copeland NG, Gilbert DJ, Jenkins NA, Lanahan AA, Worley PF (1995) Arc, a growth factor and activity-regulated gene, encodes a novel cytoskeleton-associated protein that is enriched in neuronal dendrites. Neuron 14:433–445

Macdonald PM (1990) Bicoid mRNA localization signal: phylogenetic conservation of function and RNA secondary structure. Development 110:161–171

Mandelkow E, Mandelkow E-M (1995) Microtubules and microtubule-associated proteins. Curr Opin Cell Biol 7:72–81

Marsden KM, Doll T, Ferralli J, Botteri F, Matus A (1996) Transgenic expression of embryonic MAP2 in adult mouse brain: implications for neuronal polarization. J Neurosci 16:3265–3273

Mayford M, Baranes D, Podsypanina K, Kandel ER (1996) The 3'-untranslated region of CaMKII alpha is a *cis*-acting signal for the localization and translation of mRNA in dendrites. Proc Natl Acad Sci USA 93:13250–13255

Mohr E (1999) Subcellular RNA compartmentalization. Prog Neurobiol 57:507–525

Mohr E, Richter D (1995) mRNAs in extrasomal domains of rat hypothalamic peptidergic neurons. In: Lipshitz HD (ed) Localised RNAs. Springer-Verlag, Berlin Heidelberg New York, pp 275–287

Mohr E, Morris JF, Richter D (1995) Differential subcellular mRNA targeting: deletion of a single nucleotide prevents the transport to axons but not to dendrites of rat hypothalamic magnocellular neurons. Proc Natl Acad Sci USA 92:4377–4381

Morales M, Fifkova E (1989) Distribution of MAP2 in dendritic spines and its colocalization with actin. An immunogold electron-microscope study. Cell Tissue Res 256:447–456

Morris JF, Pow DV, Sokol HW, Ward A (1993) Dendritic release of peptides from magnocellular neurons in normal rats, Brattleboro rats and mice with hereditary nephrogenic diabetes insipidus. In: Gross P, Richter D, Robertson GL (eds) Vasopressin. Libbey Eurotext, Paris, pp 171–182

Morris JF, Ma D, Pow DV, Wang H, Ward A (1997) Peptide-secreting dendrites: new controls for neuroendocrine neurons. In: Korf HW, Usadel KH (eds) Neuroendocrinology. Retrospect and perspectives. Springer, Berlin Heidelberg New York, pp 71–85

Muslimov IA, Santi E, Homel P, Perini S, Higgins D, Tiedge H (1997) RNA transport in dendrites: a *cis*-acting targeting element is contained within neuronal BC1 RNA. J Neurosci 17:4722–4733

Obar RA, Dingus J, Bayley H, Vallee RB (1989) The RII subunit of cAMP-dependent protein kinase binds to a common amino-terminal domain in microtubule-associated proteins 2A, 2B, and 2C. Neuron 3:639–645

Papandrikopoulou A, Doll T, Tucker RP, Garner CC, Matus A (1989) Embryonic MAP2 lacks the cross-linking sidearm sequences and dendritic targeting signal of adult MAP2. Nature 340:650–652

Prakash N, Fehr S, Mohr E, Richter D (1997) Dendritic localization of rat vasopressin mRNA: ultrastructural analysis and mapping of targeting elements. Eur J Neurosci 9:523–532

Quinlan EM, Halpain S (1996) Postsynaptic mechanisms for bidirectional control of MAP2 phosphorylation by glutamate receptors. Neuron 16:357–368

Ressler KJ, Sullivan SL, Buck LB (1994) Information coding in the olfactory system: evidence for a stereotyped and highly organized epitope map in the olfactory bulb. Cell 79:1245–1255

Rubino HM, Dammerman M, Shafit-Zagardo B, Erlichman J (1989) Localization and characterization of the binding site for the regulatory subunit of type II cAMP-dependent protein kinase on MAP2. Neuron 3:631–638

Schuman EM (1999) mRNA trafficking and local protein synthesis at the synapse. Neuron 23:645–648

Shafit-Zagardo B, Kalcheva N (1998) Making sense of the multiple MAP-2 transcripts and their role in the neuron. Mol Neurobiol 16:149–162

Steward O (1997) mRNA localization in neurons: a multipurpose mechanism? Neuron 18:9–12

Steward O, Halpain S (1999) Lamina-specific synaptic activation causes domain-specific alterations in dendritic immunostaining for MAP2 and CAM kinase II. J Neurosci 19:7834–7845

Steward O, Levy WB (1982) Preferential localization of polyribosomes under the base of dendritic spines in granule cells of the dentate gyrus. J Neurosci 2:284–291

Steward O, Kleiman R, Banker G (1995) Subcellular localization of mRNA in neurons. In: Lipshitz HD (ed) Localised RNAs. Springer-Verlag, Berlin Heidelberg New York, pp 235–255

Tiedge H, Brosius J (1996) Translational machinery in dendrites of hippocampal neurons in culture. J Neurosci 15:7171–7181

Tiedge H, Bloom FE, Richter D (1999) RNA, whither goest thou? Science 283:186–187

Torre ER, Steward O (1996) Protein synthesis within dendrites: glycosylation of newly synthesised proteins in dendrites of hippocampal neurons in culture. J Neurosci 16:5967–5978

Trembleau A, Morales M, Bloom FE (1996) Differential compartmentalization of vasopressin messenger RNA and neuropeptide within the rat hypothalamo-neurohypophysial axonal tracts: light and electron microscopic evidence. Neurosci 70:113–125

Tucker R, Garner CC, Matus A (1989) In situ localization of microtubule-associated protein mRNA in the developing and adult rat brain. Neuron 2:1245–1256

van Rossum D, Hanisch UK (1999) Cytoskeletal dynamics in dendritic spines: direct modulation by glutamate receptors? Trends Neurosci 22:290–295

Vassar R, Chao SK, Sitcheran R, Nuñez JM, Vosshall LB, Axel R (1994) Topographic organization of sensory projections to the olfactory bulb. Cell 79:981–991

Wensley CH, Stone DM, Baker H, Kauer JS, Margolis FL, Chikaraishi DM (1995) Olfactory marker protein mRNA is found in axons of olfactory receptor neurons. J Neurosci 15:4827–4837

Westrum LE, Jones DH, Gray EG, Barron J (1980) Microtubules, dendritic spines and spine apparatuses. Cell Tissue Res 208:171–181

Wiche G (1989) High-Mr microtubule-associated proteins: properties and functions. Biochem J 259:1–12

Wu L, Wells D, Tay J, Mendis D, Abbott MA, Barnitt A, Quinlan E, Heynen A, Fallon JR, Richter JD (1998) CPEB-mediated cytoplasmic polyadenylation and the regulation of experience-dependent translation of alpha-CaMKII mRNA at synapses. Neuron 21:1129–1139

Zuker M (1989) On finding all suboptimal foldings of an RNA molecule. Science 244:48–52

RNA Transport and Local Protein Synthesis in the Dendritic Compartment

Alejandra Gardiol, Claudia Racca, and Antoine Triller[1]

1 Introduction

The shape and composition of the postsynaptic membrane depend on synaptic activity and change throughout the lifetime of the synapse (for references, see Albright et al. 2000). The mechanisms underlying the establishment of the receptor mosaic at the postsynaptic membrane (synthesis, processing, targeting and insertion of receptors) are still poorly understood. It is currently assumed that receptors, like other transmembrane proteins, are targeted to the plasma membrane via the somatic secretory pathway. Hypotheses concerning receptor-targeting mechanisms in the central neuron derive from two different systems: the neuromuscular junction (for references, see Sanes and Lichtman 1999) and the epithelial cell as a paradigm for neuron polarity (Dotti and Simons 1990). Recently, sorting signals have been identified for the metabotropic glutamate receptors (mGluRs, Stowell and Craig 1999). However, the machinery allowing a vectorial transport of neurotransmitter receptors from the cell soma to the postsynaptic differentiation has not been identified yet. The most challenging question is: how can thousands of signals, which notify minute changes occurring at each postsynaptic differentiation and arise from a large number of synapses, be treated somatically and subsequently induce a synaptic change, some synapses being located on dendrites hundreds of microns away? A model has been postulated in which the rough structure of the postsynaptic membrane is determined by a classical somatic post-Golgian targeting of receptors and the local and individual fine regulation of the postsynaptic membrane composition is achieved at the synaptic level. Experimental data suggest that, on physiological stimulation, new receptors can be inserted into the postsynaptic membrane (Quinlan et al. 1999). These receptors can either be targeted from a ready-to-use subsynaptic pool (Maletic-Savatic et al. 1998; Maletic-Savatic and Malinow 1998; Nusser et al. 1998; Luscher et al. 1999; Rubio and Wenthold 1999; Shi et al. 1999; Racca et al. 2000) or translated in situ from localized subsynaptic mRNAs (Steward 1997; Racca et al. 1997a; Gardiol et al. 1999; Quinlan et al. 1999). These two mecha-

[1] Laboratoire de Biologie Cellulaire de la Synapse N&P INSERM U497 Ecole Normale Supérieure, 46 rue d'Ulm, 75005 Paris, France
Present address: C. Racca, Laboratory of Neurochemistry, NIDCD-NIH, Bethesda, MD 20892, USA

Results and Problems in Cell Differentiation, Vol. 34
D. Richter (Ed.): Cell Polarity and Subcellular RNA Localization
© Springer-Verlag Berlin Heidelberg 2001

nisms are not exclusive, since the subsynaptic pool can either be targeted from the somatic secretory apparatus or synthesized in situ from localized mRNAs. Locally synthesized receptors could then be stored before being inserted into the postsynaptic membrane. During the last decade, convergent evidence supports the notion of local, restricted translation of proteins from perisynaptically located mRNAs. In many non-neuronal systems, restriction of protein synthesis to a subdomain contributes to creation and maintenance of cell or organism polarization (for references, see Bassell et al. 1999; Lasko 1999; Mowry and Cote 1999), avoids the mislocalization of a protein (e.g. myelin basic protein, MBP; Ainger et al. 1993) and enables the in situ delivering/renewing of a protein (e.g. actin/fibroblasts; Lawrence and Singer 1986). In neurons, the mechanisms which contribute to ontogenesis in oocytes and in embryos may also be involved in remodeling and strengthening of synapses (e.g. Kang and Schuman 1996; Quinlan et al. 1999). In this chapter, we will focus on works which sustain the hypothesis of the autonomy of the synapse, a concept originally proposed by Steward and coworkers (Steward and Falk 1986; Steward 1987, 1997).

2 Differential Subcellular RNA Localization in Neurons

2.1 Dendritic RNAs

Neurons are highly polarized cells, which can be subdivided into three different specialized domains: soma, dendrites, and axons. Each of these main domains can be further subdivided into smaller domains (e.g. growth cones, synaptic differentiations, spines, Ranvier nodes) with highly specialized functions. With progress in detection procedures, molecules once believed to be restricted to the neuronal soma have also been found in other cellular domains.

The first mRNA to be found in dendrites was that for MAP-2. It was detected by radioactive in situ hybridization (ISH) in dendrites of pyramidal cells in the molecular layer of the hippocampus, as well as in the neocortex of developing rat brain (Garner et al. 1988). This discovery was particularly significant because MAP-2 is a dendritic-specific protein promoting tubulin polymerization and participating in dendrite morphogenesis. In young cultured hippocampal neurons (5 DIV), various poly(A) RNAs, rRNAs and MAP-2 mRNAs are present in all cellular compartments, soma and neurites, as they still have not differentiated to become dendrites and axons. In contrast, mRNAs coding for GAP-43, β-tubulin and β-actin never translocate in neurites even at early undifferentiated stages (Bruckenstein et al. 1990; Kleiman et al. 1990, 1993a, 1994), at least in hippocampal and sympathetic neurons (see Bassell et al. 1998). Therefore, mechanisms controlling the restriction of subclasses of mRNAs to the soma are already operant in neurons at the early stages. This suggests that mRNA compartmentalization may contribute to the establishment of neuronal polarization. All dendritic proteins do not benefit from dendritic localization of their corresponding mRNAs, suggesting that mRNA compartmentalization is not a default pathway, but rather a finely regulated mechanism. A striking example is provided by the dendritic exclusion or localization of the mRNAs coding for the different subunits of the calcium-calmodulin-dependent protein kinase II (CaMKII; Burgin et al. 1990). This multimeric holoenzyme is composed of an array of α and β subunits (Bennett et al. 1983). The expression and stoichiometry of the enzyme vary during the different developmental stages. In the hippocampus and neocortex, the mRNAs encoding the α subunits are present in dendrites, whereas those coding for the β subunits are only detected in the somata of the same cells. Both mRNAs are excluded from the dendrites of cerebellar Purkinje cells (Burgin et al. 1990). Moreover, the developmental changes in the protein amount are correlated with changes in the mRNA content, suggesting that transcription may regulate the levels of CaMKII during brain development. At the ultrastructural level, mRNAs encoding the α subunit of CaMKII were found in apical dendrites of pyramidal neurons of the hippocampus (Martone et al. 1996). This localization of CaMKII mRNA is of paramount importance for the following reasons: (1) the enzyme CaMKII constitutes 40% of the protein content of the postsynaptic densities (PSDs; Kelly et al. 1984); (2) changes in CaMKII content can be correlated with long-term potentiation (LTP; Silva et al. 1992a, b; Ouyang et al. 1999; Steward and Halpain 1999); and (3) the time course of CaMKII mRNA content correlates with synaptogenesis (Burgin et al. 1990).

2.2 Synaptic mRNAs

Some dendritic mRNAs code for proteins involved directly or indirectly in synaptic function. Some of these mRNAs, even if not directly detected under synapses, can be considered as synaptic mRNAs because: (1) the proteins they encode have synapse-exclusive functions (e.g. neurotransmitter receptors) and (2) their amounts are modified by activity or they can play a role in postsynaptic signaling.

2.2.1 Neurotransmitter Receptor mRNAs

The morphological demonstration of synapse-associated mRNAs is impaired by technical limitations. Our group was the first to demonstrate a subsynaptic localization of mRNAs coding for a neurotransmitter receptor (Racca et al. 1997a). In the spinal cord of adult rats, where glycinergic inhibition predominates, we used alkaline phosphatase-ISH to demonstrate a somato-dendritic localization of glycine receptor (GlyR) α1 and α2 subunit mRNAs. In contrast, mRNAs encoding the β subunit and the GlyR-associated protein, gephyrin, were only detected in the soma. Electron microscopy and horseradish peroxidase- or immunogold-ISH allowed more precise detection of mRNAs coding for the α subunits under synapses, where they are associated with subsynaptic cisternae. Since GlyR clusters accumulate in front of inhibitory boutons (Triller et al. 1985), it could be assumed that these subsynaptic mRNAs may participate in a local subsynaptic synthesis of α subunits. GlyR can exist as an αβ hetero-oligomer, anchored to the cytoskeleton by GlyRβ subunit-gephyrin interaction (Meyer et al. 1995) or as an α homomer (Hoch et al. 1989; Takahashi et al. 1992; Betz et al. 1999). The failure to detect β subunit and gephyrin mRNAs under synapses raises the question of GlyR cluster composition. Several possibilities may account for lack of detection of these mRNAs. First, the amounts or stability of β subunit and gephyrin mRNAs could be lower in dendrites and under synapses, thus precluding their detection. Second, since oligomeric receptor subunits have to be assembled before their exit from the endoplasmic reticulum (ER; for references, see Green and Millar 1995) it could be that the β subunit protein is already present in the subsynaptic ER (see below). The β subunit could subsequently associate with the α subunit translated from the synapse-associated mRNAs. Hence, gephyrin could recruit this newly formed GlyR to the nearby postsynaptic sites. Finally, in zebra fish, electrophysiological approaches favor the notion that functional homomeric GlyRα receptors (Schmieden et al. 1989) could be synaptic (Legendre 1997).

Other mRNAs encoding neurotransmitter receptors were found in dendrites. In cultured hippocampal neurons, Miyashiro and coworkers (1994) showed the presence of mRNAs coding for different members of the glutamate receptor (GluR) family by using RNA amplification techniques applied to the cytoplasm of a single isolated dendrite. In another study, the mRNAs encoding the 1α subtype of the metabotropic GluR (mGluR1α) were detected by radioactive ISH in dendrites of the fusiform cells of the cochlear nucleus (Bilak and Morest 1998). mRNAs coding for the NR1 subunit of the *N*-methyl-D-aspartate receptor (NMDAR) are transported into dendrites of hippocampal pyramidal cells during the neo-synaptogenesis and axonal sprouting which follow the unilateral transection of the perforant path (Gazzaley et al. 1997). The presence within dendrites of mRNAs coding for neurotransmitter receptor subunits, even if the resolution of the methods does not allow their ultrastructural detection, strongly suggests that their dendritic localization reflects a synaptic targeting of these mRNAs.

2.2.2 Non-Neurotransmitter Receptor Synaptic mRNAs

Some RNAs which are not directly implicated in synaptic function are nonetheless directed to specific dendritic domains upon synaptic stimulation.

The mRNAs coding for the immediate early gene Arc/arg3.1, for activity-regulated cytoskeleton-associated protein, whose expression is induced by growth factor stimulation, translocate to dendrites of the molecular layer of the hippocampus following synaptic activation mediated by NMDA receptors (Link et al. 1995; Lyford et al. 1995; Steward et al. 1998b). The Arc mRNA encodes a soluble protein associated with the cytoskeleton whose function is still unknown (Link et al. 1995; Lyford et al. 1995).

mRNAs encoding the brain-derived neurotrophic factor (BDNF), and its receptor, the tyrosine receptor kinase B (TrkB), were observed in dendrites of cultured hippocampal neurons. Furthermore, TrkB and BDNF mRNAs are transported distally in dendrites following synaptic activation (Tongiorgi et al. 1997). This activity-dependent mRNA transport may underlie the subsynaptic protein synthesis of BDNF and TrkB (see below). It was shown that neurotrophins can be secreted by dendrites and that the TrkB receptor is expressed in the dendritic membrane (Blochl and Thoenen 1996). The ligand-receptor complex (BDNF-TrkB) is probably involved in the modulation of synaptic transmission and in hippocampal LTP (McAllister et al. 1999)

The non-coding RNA BC1 is also present in hippocampal dendrites (Tiedge et al. 1991). In cultured hippocampal neurons, labeling appears to be regulated by synaptic activity (Muslimov et al. 1998). The function of BC1 RNA is still not well understood, but phylogenetic studies suggest that it may participate in the regulation of the translation of transported mRNA (DeChiara and Brosius 1987; Skryabin et al. 1998; Brosius 1999)

2.3 Axonal and Presynaptic mRNAs

Synaptically-targeted mRNAs have also been studied in invertebrates where neurons do not have clear-cut dendritic or axonal compartments. Cultures of *Aplysia* neurons are very useful systems for studying transport of mRNAs as the pre- and postsynaptic neurons can be selectively activated. Furthermore, postsynaptic targets of the same presynaptic neuron with different properties can be identified. The mechanosensory neurons of *Aplysia* contact two different target cells: one (L7) with which it establishes a synapse, and a second with which it fails to induce the formation of a synaptic contact. In this model it was shown that mRNA encoding for sensorine A, a peptide involved in synapse formation, migrated only in the branch contacting the L7 target cell, suggesting that mRNA transport mechanisms may participate in synapse establishment (Schacher et al. 1990). Furthermore, local protein synthesis in the presynaptic element is required for long-term facilitation (LTF) in cultured *Aplysia* neurons (Martin et al. 1997). These results cannot be simply extrapolated to vertebrates, since in *Aplysia* the compartment affected by RNA transport and protein synthesis is the presynaptic one. However, these results emphasize the fact that mRNA transport and local protein synthesis can be key steps in the establishment of synaptic contacts.

In vertebrates, RNAs are also located in axonal compartments, principally in cells belonging to the neuro-hormonal system. These mRNAs often code for hormones secreted in the blood circulation by axonal swellings. Since the ribosomes and other members of the synthetic apparatus have not yet been detected in the axonal tract, the role of these RNAs is speculative. Several hypotheses have been put forward concerning this compartmentalization. The localization could be non-specific, or mRNAs could be directed to the axons to be stored or degraded (for references, see Mohr 1999). The non-coding RNA

BC1 detected in hippocampal dendrites was also shown to be present in axons of rat hypothalamic cells (Tiedge et al. 1993).

3 RNA Transport

The first evidence that mRNAs can move into dendrites emerged from experiments in which ^{3}H-uridine incorporated into neosynthesized RNA could be detected in dendrites of cultured hippocampal neurons (Davis et al. 1987). This transport was found to be energy-dependent and to rely on the integrity of the cytoskeletal network. Later, different techniques (time-lapse microscopy, vital nucleic markers, tagged RNAs . . .), as well as data on other cellular or embryonic models (*Drosophila*, *Xenopus*, yeast; see Bassell et al. 1999; Etkin and Lipshitz 1999; Jansen 1999; Lasko 1999; Mowry and Cote 1999) helped to develop some general concepts underlying the cell biology of RNA transport in neurons. The RNA movements are: (1) microtubule-dependent (Davis et al. 1987; Knowles et al. 1996; Bassell et al. 1998); (2) have a rate of movement that ranges between 20 and 400 µm/h (Davis et al. 1987; Knowles et al. 1996; Muslimov et al. 1997; Bassell et al. 1998; Wallace et al. 1998); (3) do not depend on protein synthesis (Davis et al. 1987; Kleiman et al. 1993b; Steward et al. 1998a; Wallace et al. 1998); (4) are energy-dependent (Davis et al. 1987); (5) do not depend on RNA diffusion, but on the association of RNA with proteins in ribonucleoparticles (RBPs, Wilhelm and Vale 1993; Kohrmann et al. 1999). Thus, some protein factors (RNA-binding proteins, cellular motors, proteins implicated in the translation), called *trans*-acting factors, may bind the mRNAs at specific sequences, called *cis*-elements, to build macromolecular complexes that deliver the RNAs to their destinations.

3.1 *trans*-Acting Factors

RNAs move along microtubules as macromolecular particles, also referred to as granules. Several constituents of these particles have been identified in different cell types (see Kiebler and DesGroseillers 2000). The constituents of the particle can be classified into three groups: (1) those belonging to the pathway of translation, (2) those allowing the movements of the particle, and (3) those ensuring the recognition of RNAs that have to be transported.

3.1.1 Translational Factors and Transport Granules

In oligodendrocytes, the MBP mRNA moves into the processes along microtubules as macromolecular complexes containing arginyl-tRNA synthetase, elongation factor 1α, and ribosomal RNAs (Ainger et al. 1993; Barbarese et al. 1995). In neurons, granules contain the 60 S ribosomal subunit, as well as the elongation factor 1α (Knowles et al. 1996; Bassell et al. 1998). Furthermore, other constituents of the synthetic machinery were shown to be present within dendrites (see below).

The question of the interaction between mRNA transport and translation is central to understanding the relationship between localized protein synthesis and activity. Two sets of data help to solve the problem. First, transport of mRNAs occurs even if protein synthesis is inhibited (Davis et al. 1987; Kleiman et al. 1993b; Steward et al. 1998a; Wallace et al. 1998). The simplest model postulates that mRNA within these particles can only be translated when it reaches its appropriate docking locus. Second, mRNAs that have a soma-restricted distribution (e.g. ferritin, tubulin, and GAP43) start to translocate to dendrites when translation is inhibited (Kleiman et al. 1993b; Lu et al. 1998). If ribosomes were responsible for the trapping of somatic mRNAs, the presence of mRNAs in dendrites would result from an overflow after release from ribosomes. This hypothesis is in contradiction with the fact that transport granules contain ribosomal elements. The fact that protein synthesis inhibition alters the distribution pattern of mRNAs, normally restricted to the soma, is in favor of the notion that the inhibition of protein synthesis stops the generation of rapid-turnover molecules responsible for the trapping of somatic mRNAs (Kleiman et al. 1993b).

3.1.2 Granule Motors

Trans-acting factors should also include motors responsible for mRNA translocation. Although it has been established that mRNA transport is energy-dependent (Davis et al. 1987), data on dendritic motors for RNA movements are indirect. The speed of RNA movements varies widely (20–400 µm/h, see above for references). Differences between measured values can be largely attributed to the methods employed (pulse-chase versus time-lapse experiments, or different cellular systems). Moreover, the values are often vectorial means of the speeds, and in several time-lapse experiments it was observed that a single particle displays complex patterns of movements (retrograde, anterograde, stochastic; Ainger et al. 1993; Knowles et al. 1996). These differences suggest that the RNA motors are different from the known motors involved in axonal transport. Until now, no specific proteins responsible for mRNA motility have been identified in neurons. In yeast, a putative RNA-motor (SHE1) with sequence similarity to a myosin V motor was identified in a yeast mutant (Bertrand et al. 1998). Moreover, the speed of RNA particle movements in wild-type yeasts was in the range of myosin-driven actin-based movements. This is in apparent contradiction with the body of knowledge from neurons where the bulk of the data suggests that RNAs move along microtubules. This is not a general rule, since in some other eukaryotic systems (fibroblasts, muscle cells, for example), the β-actin mRNA moves along microfilaments (Singer et al. 1989; Sundell and Singer 1991).

3.1.3 mRNA-Sspecific trans-Acting Factors

Since RNA transport is RNA-specific, the transport particles should contain molecules ensuring this specificity. A protein, Vera, has been identified in *Xenopus* embryos using UV cross-linking (Deshler et al. 1997). During *Xenopus* embryogenesis, Vera binds the Vg1 mRNA and its transport is microtubule-dependent. Moreover, Vera colocalizes with the ER and coprecipitates with TRAPα, a constituent of the signal recognition particle. It was then suggested that Vera was responsible for Vg1 mRNA localization through an interaction with the ER. In the *Drosophila* oocyte, the RNA-binding protein Staufen is required for the proper localization of two maternal mRNAs involved in the anteroposterior polarization of the fly oocyte (St. Johnston and Nusslein-Volhard 1992; Ferrandon et al. 1994, 1997). Recently, human, mouse (Marion et al. 1999; Wickham et al. 1999), and rat (Kiebler et al. 1999) homologues of *Drosophila* Staufen have been cloned. Staufen homologues were found by Western blotting in most human tissues, including the brain. The vertebrate sequences are homologues to the *Drosophila* one. Human and mouse fusion proteins bind in vitro double-stranded RNAs as well as tubulin (Wickham et al. 1999). Interestingly, the mammalian Staufen associates with microtubules, polyribosomes and the ER and presents a tubulin binding site homologous to the MAP1 tubulin binding domain (Wickham et al. 1999). In neurons, Staufen immunoreactivity has been detected in the somato-dendritic compartment of cultured hippocampal cells, near the postsynaptic differentiations and in close association with microtubules (Kiebler et al. 1999). In living neurons transfected with a GFP-tagged Staufen construct, Kohrmann and coworkers (1999) showed that Staufen forms large RNPs that move into dendrites, along microtubules in a non-vesicular pathway. However, in order to elucidate the role of Staufen in RNA transport and/or targeting, the mammalian RNA partner of Staufen has to be identified. Another neuron-specific protein belonging to the Staufen family, the visinin-like protein (VILIP) was recently identified by PCR amplification of brain-derived cDNAs (Mathisen et al. 1999). The RNA binding to VILIP is calcium-dependent and may link RNA transport to calcium entry and therefore to synaptic activity.

Another family of RNA-binding proteins, with a KH domain, involved in RNA splicing, is thought to participate in dendritic transport (Burd and Dreyfuss 1994; Darnell 1996). Interestingly, some of these KH domain family proteins are also incriminated in neuronal diseases. This is the case for fragile X mental retardation protein (FMRP) and Nova (novel ventral neuron-specific antigene). Both proteins present KH domains (homologous to the RNA-binding domain of the hnRNP protein K; Siomi et al. 1993) bind their own mRNA, and present nuclear localization sequences. Fragile X syndrome is linked to mutations in the FMR1 gene encoding FMRP and leads to mental retardation, macrorchidism and facial abnormalities (De Boulle et al. 1993) Mutations that abolish the mRNA-FMRP interactions at the KH domain generate severe forms of these hereditary diseases (Siomi et al. 1994). FMRP mRNA localizes in the neuronal nucleoplasm as well as in the somato-dendritic compartment where it forms RNPs in association with ribosomes (Feng et al. 1997a, b; Weiler et al. 1997). Furthermore, FMRP mRNA is locally translated near synapses in response to stimulation by mGluR agonists (Weiler et al. 1997). Nova-1, a neuron-specific RNA-binding protein belonging to the KH family, is involved in POMA (paraneoplastic opsoclonus myoclonus and ataxia) a clinical syndrome associated with a specific set of cancers (see Darnell 1996; Posner and Dalmau 1997). In addition, Nova-1, which binds to the GlyR $\alpha 2$ subunit messengers, is detected in the somatic and dendritic cytoplasms of the spinal cord and brain stem neurons (Buckanovich et al. 1993) where it forms RNPs (Racca et al. 1997b). Hence, Nova proteins may be key factors that specifically recognize some synaptic RNAs, thereby allowing their incorporation in RNPs which will then be transported. However, the role of the KH proteins is more complex since they are also implicated in the processing of immature mRNAs (Burd and Dreyfuss 1994). Therefore, new metabolic pathways linking RNA splicing and RNA transport might account for local protein synthesis at synapses.

3.2 *cis*-Acting Sequences

The localization of a given RNA seems to be based on a complex interplay between the *trans*-acting factors, the cytoskeleton, and targeting *cis*-acting sequences of mRNAs. In the soma, mRNAs encoding secreted or membrane proteins are targeted to the ER by specific amino acid sequences on the nascent peptide chain (signal peptide; for references, see Pelham and Munro 1993). The dendritic targeting of mRNAs seems to involve a different mechanism as mRNAs can be transported even if translation is blocked (Steward et al. 1998a; Wallace et al. 1998).

In neurons, these dendritic-targeting elements are often located in the 3'UTR regions of mRNAs (CamKIIα, Mayford et al. 1996b); MAP-2 (Blichenberg et al. 1999). This is also the case in other systems (*Xenopus* oocytes, *Drosophila* oocytes or embryos, fibroblasts; see Decker and Parker 1995). The polymerase III product, BC1, does not have a 3'UTR. Nevertheless, BC1 also presents 62 nucleotides in its 3' region that account for its dendritic transport (Muslimov et al. 1997). Recently, Blichenberg et al. (1999) have demonstrated, by transfecting or microinjecting different constructs in cultured neurons, that 640 nucleotides belonging to the 3'UTR, are both sufficient and necessary for the localization of MAP-2 mRNAs. Conversely, in some instances, the dendritic targeting elements were found to be located in the coding region (Kindler et al. 1996) or in a region comprising the coding region and the 3'UTR (Prakash et al. 1997). Surprisingly, the neuronal targeting elements do not share homologous sequences between themselves or with *cis*-acting elements of other cell-type RNAs. Altogether, these results support the idea that since primary nucleotide sequences of these regions are different, the localization signal must lay in the secondary or tertiary structures assumed by the mRNAs and recognized by the transport apparatus.

Besides an involvement in mRNA transport, the 3'UTR region is also involved in the regulation of the stability and translation of mRNAs (see Decker and Parker 1995). Therefore, the 3'UTR region of mRNAs is a complex box interacting with the *trans*-acting factors, hence modulating the metabolism, localization and translation of the dendritic and synaptic mRNAs. The understanding of the regulation of mRNA transit is one of the missing links relating synaptic activity and RNA metabolism. Recently, some of the steps linking stability to the transport of mRNA as a function of synaptic activity have been identified (Wu et al. 1998; see below).

4 Synapse and Local Protein Synthesis Machinery

4.1 Ultrastructural Data Suggesting Subsynaptic Protein Synthesis

Before MAP-2 and other mRNAs were detected in hippocampal dendrites (1988), several studies analyzed the ribosomal content of the post-synaptic region of hippocampal synapses. Groups of polyribosomes were found beneath the bases or in the neck of the dendritic spines, sometimes associated with cisternae, and less frequently within the spine cytoplasm, but not associated with the spine apparatus (Steward and Levy 1982; Steward and Falk 1985, 1986). The presence of ribosomes under synapses and the increase in ribosome abundance, as synaptogenesis takes place in the hippocampus, was in favor of local protein synthesis regulated by the activity (Steward and Falk 1985). In addition, several cisternae of various sizes are present close to synapses. Depending on the neuronal type, and their shape, some of these cisternae were called hippolemal cisternae (Kaiserman-Abramof and Palay 1969), subsurface cisternae (Rosenbluth 1962), and subsynaptic cisternae (Peters et al. 1991). Most of the time, ribosomes were not detected in association with these membranous elements. Rather, this is why they have been related to the smooth endoplasmic reticulum (SER). By analogy with the highly specialized sarcoplasmic reticulum of the muscle cell, these neuronal endomembranes were often thought to be related to calcium homeostasis (Rossier and Putney 1991; Villa et al. 1991; Takei et al. 1992; Pozzo-Miller et al. 1997). However, in some cases, the latter were suspected to be involved in protein synthesis (Bodian 1965; Pozzo-Miller et al. 1997). Moreover, the endoplasmic reticulum of spine apparatus, connected to the dendritic shaft one, was thought to undergo exocytosis and to contain part of the translocation machinery (Spacek and Harris 1997; Pierce et al. 2000).

4.2 Immunocytochemical Characterization of the Dendritic Translational Machinery

In the classical view, the rough endoplasmic reticulum (RER) and the Golgi apparatus (GA) are believed to extend from the soma only within the proximal portions of dendrites (Peters et al. 1991). These descriptions were based on the ultrastructural morphology of these organelles. However, immunocytochemistry and light microscopy analysis of rat brain tissue sections showed the presence of threads immunoreactive for GA markers which run into dendrites for some distance (De Camilli et al. 1986). In cultured neocortical neurons, the distribution of GA during development was analyzed with the *trans*-Golgi network marker TGN38 (Lowenstein et al. 1994). A few hours after plating, at the time when neurons start to differentiate, the TGN38 immunoreactivity (IR) was detected in several vesicle-like structures dispersed through the cytoplasm. As the neurons differentiate, the IR structures tend to fuse and the TGN38 IR acquires a classical pattern penetrating into dendrites, often the largest ones. Following our finding that GlyRα subunit mRNAs accumulated under synapses, we have reinvestigated the organization of the synthesis and secretion machinery in neuronal dendrites. The use of various markers, Rab1 (Saraste et al. 1995), CTR433 (Jasmin et al. 1989) and TGN38 (Wilde et al. 1992), revealed that the subcellular distribution of GA in spinal cord neurons was more complex than initially suspected (Gardiol et al. 1999). High-resolution confocal images revealed a thread-like central GA connected to the somatic stacks and invading some dendrites. Besides this classical pattern, we detected a punctiform and discontinuous labeling localized in front of synaptic boutons even in dendrites devoid of the central GA thread. This punctiform pattern is reminiscent of that found by Lowenstein and coworkers (1994) in developing dendrites as well as that detected around fundamental nuclei in the muscle cell (Jasmin et al. 1989, 1995). Therefore, besides the classical GA that is formed by multicisternal stacks connected with the somatic one, other smaller elements fulfilling Golgi functions may exist near synaptic contacts. Ultrastructurally, these subsynaptic elements were minute cisterna-like structures of different shapes and sizes which were indeed immunopositive for the different GA markers (Gardiol et al. 1999). Moreover, we also found a set of subsynaptic membranous structures positive for ER, reticulum markers Bip (Bole et al. 1986) and ribosomal-P-proteins (Elkon et al. 1985). Therefore, some of these subsynaptic structures have one feature of rough ER. These results suggest that a miniature synthesis and secretion machinery may exist under the synapse and strongly favor the notion of metabolic autonomy of the synapse.

In cultured hippocampal neurons, several components of the translational machinery have been reported. Tiedge and Brosius (1996) demonstrated the presence in dendrites of tRNAs, aminoacyl-tRNA-synthetases as well as ribosomes. They also showed that initiation (eIF2) as well as elongation factors (eEF2) were available in dendrites. Initiation factors of the translation had also been detected ultrastructurally, associated with subsynaptic cisternae thought to be involved in protein synthesis (Gardiol et al. 1999). Altogether, these results suggest not only that subsynaptic translation might be achieved but also that it may be regulated (see Rhoads 1993, and below).

Some components of the synthetic machinery have not yet been detected in the dendritic domain, for example the TRAPα protein, as well as 7SL RNA, both components of the signal recognition system associated with mRNAs encoding transmembrane proteins (Tiedge and Brosius 1996).

4.3 Biochemical Studies of Translation and Glycosylation Within Dendrites

Several biochemical studies have emphasized the translational capacities of dendrites. Synaptosome preparations from the mossy-fiber/CA3 synapse region have revealed the presence of RNA populations (CaMKII, BC1; Chicurel et al. 1993). Moreover, it has been shown that the cytoplasm of spines of the CA3 region contains large amounts of ribosomes (Chicurel and Harris 1992; Chicurel et al. 1993), and that synaptosomal preparations can perform protein synthesis (Rao and Steward 1991).

As an alternative to the use of synaptosomes, a preparation of metabolically active dendrites from hippocampal cultured neurons was developed (Torre and Steward 1992). With this system, dendritic protein synthesis could be demonstrated by the puromycin-sensitive, chloramphenicol-resistant incorporation of 3H-leucine in processes subsequently identified as dendrites by their MAP2-IR. The post-translational modifications following protein synthesis could also be established with the same model by combining immunocytochemistry and autoradiographic analysis of incorporation of radiolabeled precursors of glycosylation (Torre and Steward 1996). This work demonstrated that (1) glycosylation machinery is physically present in dendrites and (2) glycosylation can be performed in a protein synthesis-dependent manner in dendritic segments.

5 Activity-Dependent mRNA Transport and Protein Synthesis

A 'trophic function' of subsynaptic polyribosomes and mRNAs was postulated by Bodian some years ago (1965). It is only recently that translation from synaptically localized mRNAs has been demonstrated. Signaling pathways that may underlie these phenomena are now beginning to be unraveled. The rationale is that neurotransmitter release and binding to postsynaptic receptors should trigger the amplification of this primary stimulus, which finally promotes long-lasting structural and functional modifications in the postsynaptic cell. For instance, the unraveled steps include: (1) calcium entry following neurotransmitter binding by postsynaptic receptors, (2) activation of postsynaptic calcium sensitive effectors (Calmodulin, CaMKII, CREB), and (3) activation of second effectors participating in complex phosphorylation cascades (MAP-2, eEF2, CPEB).

5.1 Glutamate Receptor Activation and Calcium Entry

Local synthesis of the NR2A subunit of the NMDAR has been proposed to account for the rapid expression of new NMDARs at neocortical synapses, when visually deprived rats undergo visual experience (Quinlan et al. 1999). This has led to the suggestion that both local synthesis of new NMDAR subunits in Golgi-like structures in spines and their subsequent insertion into the plasma membrane can take place. This neo-insertion allows a rapid change in NMDA receptor subunit composition in response to experience. Indeed, mRNAs encoding NMDAR subunits could already be present beneath synapses or be routed to the synapse after synaptic activation and/or during synapto-genesis. Interestingly, during post-lesional neosynaptogenesis, the NR1 subunit mRNA of NMDAR is targeted toward synapses and is locally translated (Gazzaley et al. 1997). A direct implication of NMDA receptor transmission and Ca^{2+} in local protein synthesis of CaMKII was recently demonstrated in experiments performed in synaptosomes or in vivo (Quinlan et al. 1999; Scheetz et al. 2000). This primary entry of calcium triggers complex events: (1) cascades of phosphorylations through calcium-dependent kinases and (2) secondary calcium release from intracellular stores (Llano et al. 1994). These two events are counteracted by the activation of phosphatases and calcium sequestration. Hence, calcium waves and phosphorylations/dephosphorylations may be the source of cellular oscillators regulating mRNA transport and local protein synthesis. NMDA-dependent calcium entry also regulates transcription by inhibiting the elongation step via the phosphorylation of elongation factors (Ryazanov et al. 1988; Prostko et al. 1995; Srivastava et al. 1995). The role of AMPA and NMDA receptors in hippocampal LTP is well established. Interestingly, mGluR activation increases the aggregation of polyribosomes (Weiler and Greenough 1993) and the translation at the synapse of mRNA via phosphatidylinositol hydrolysis (FMRP, Weiler et al. 1997). Phospholipid hydrolysis and PKC activation following mGluR stimulation enhance polyribosomal loading. This effect is inhibited if NMDA agonists or calcium ionophores are applied to the synaptosomal preparation prior to mGluR activation. This may involve the inactivation of phospholipase A2 by CamKII (Weiler et al. 1996). NMDA blockers applied after mGluR stimulation do not abolish the induced polyribosomal aggregation, suggesting a role for intracellular Ca^{2+} stores (Weiler and Greenough 1993). Opposite effects triggered by delayed activation of different sets of receptors at the synaptic differentiation could therefore explain some apparent contradictions, such as the enhancement of translation rate of certain messengers (e.g. CaMKIIα subunit) as general translation is inhibited (Marin et al. 1997; Scheetz et al. 1997, 2000). Messengers coding for GluRs (Miyashiro et al. 1994) and for inositol 1,4,5-triphosphate type1 receptors (InsP3R1, endoplasmic receptor responsible for the opening of intracellular Ca^{2+} stocks, Furuichi et al. 1993) have been shown to be dendritic. Therefore, the amount of ionoropic and metabotropic receptors can be locally regulated as a function of activity. Furthermore, NMDAR antagonists and calcium chelators were shown to impair RNA movements triggered by synaptic activation (Link et al. 1995). Calcium might regulate mRNA transport toward synapses via calcium-sensitive RNA-binding proteins (Mathisen et al. 1999). Finally, calcium gradients also regulate channel density at the postsynaptic membrane, as well as the stability of channel mRNAs (Schorge et al. 1999), and the spine shape (Korkotian and Segal 1999; Segal et al. 2000).

5.2 Ca^{2+}-CaMKII Activation and Phosphorylation Pathways

CamKII, the most abundant component of the postsynaptic differentiation, is a multimeric holoenzyme activated by the binding of the calcium-calmodulin complex. The activated enzyme undergoes a rapid autophosphorylation which enables its association with the PSD and amplifies its own activity by diminishing the calmodulin dissociation rate. Transient entry of calcium prolongs the kinase activity, since the enzyme catalyzes its own phosphorylation. CamKII substrates are intimately involved in the synaptic function. This enzyme catalyzes the phosphorylation of the AMPA receptor, thereby stabilizing a high conductance state. Both calcium-independent CamKII transgenic mice and AMPA GluR1-subunit knockout mice lack hippocampal LTP (Mayford et al. 1996a; Zamanillo et al. 1999). The link between CaMKII and synaptic remodeling remains obscure, but recent work has helped to unravel some metabolic pathways. In *Caenorhabditis elegans*, CamKII is implicated both in GLR1 trafficking and stabilization at the synapse, where the phosphorylation of GLR1 is involved (Rongo and Kaplan 1999). Consistent with this view, CamKII plays a key role in calcium-evoked dendritic exocytosis (Maletic-Savatic and Malinow 1998; Maletic-Savatic et al. 1998). Furthermore, the inhibition of the endogenous NSF-SNAP interaction blocks the postsynaptically-regulated exocytic pathway and abolishes LTP (Lledo et al. 1998). Therefore, new receptor subunits could be secreted and inserted at the synapse as a result of synaptic activity, in part through calcium-dependent activation of CamKII. This neo-insertion could follow a neo-translation at the synapse, as proposed by Quinlan and coworkers (1999).

The role of CamKII in synaptic plasticity is also pivotal since its protein kinase function and mRNA stability and translation are regulated by synaptic activity. Translation of CamKIIα subunit mRNA at the synapse is enhanced by NMDA receptor activation (Scheetz et al. 2000). The latter involves the phosphorylation of eukaryote elongation factor 2 (Scheetz et al. 2000), shown to be present beneath synapses (Gardiol et al. 1999). This factor is phosphorylated by another specific calcium calmodulin-dependent protein kinase different from CamKII (Ryazanov 1987; Ryazanov et al. 1988; Gardiol et al. 1999). As mentioned before, the role of the elongation factor in synaptic protein synthesis is paradoxical since its phosphorylation inhibits general protein synthesis but increases the translation of the CamKIIα subunit mRNA (Marin et al. 1997; Scheetz et al. 1997, 2000). Moreover, the Ca^{2+}-dependent phosphorylation of eukaryotic initiation factor 2 (eIF2) by the double-stranded RNA-dependent protein kinase (PKR) also inhibits the translation by stopping the initiation step (Prostko et al. 1995; Srivastava et al. 1995). Various mechanisms accounting for this paradox have been proposed, yet the general meaning of this phenomenon remains unclear. In addition to in vitro experiments showing the synaptic translation of the CamKIIα subunit messenger, in vivo experiments demonstrate that the visual experience in light-deprived rats (Wu et al. 1998) enhances the translation rate of the messenger. Furthermore, the authors show that CamKII α subunit mRNA translational rate and stability could be regulated by the cytoplasmic polyadenylation element binding protein (CPEB). This CPEB-mediated cytoplasmic polyadenylation of messengers harboring cytoplasmic polyadenylation elements (CPE) can prompt them in their synaptic translation at the expense of other non-polyA-tagged mRNAs. This switch might be achieved by the release of synaptic ribosomes engaged in another translation by stopping the initiation and elongation steps. These recent findings open up new lines of research focusing on the regulation steps allowing the synaptic translation of specific mRNAs.

5.3 The Neurotrophin Signaling Pathway

Neurotrophins (NGF, BDNF, NT-3, NT-4/5) and their tyrosine kinase-coupled receptors (TrK) are known to be involved in neuronal survival. They also mediate some forms of synaptic plasticity. Neurotrophins and their receptors are widely expressed in the CNS, particularly in those regions involved in neuronal growth and differentiation during development and in regions of the adult animal undergoing activity-dependent modifications (see Barbacid 1994). Acute application of NT-3 and BDNF to hippocampal slices stimulates protein synthesis and at the same time potentiates synaptic transmission at the Schaffer collateral/CA1 synapses (Kang and Schuman 1996). Consistent with this view, neurotrophins have been shown to increase postsynaptic ion channel density (Sharma et al. 1993; Timpe and Fantl 1994; Lesser and Lo 1995; Bowlby et al. 1997). Membrane excitability is also modified by neurotrophins via other pathways, such as the phosphorylation of NMDA receptors (Suen et al. 1997), the modification of channel properties (Jarvis et al. 1997), and the increase in intracellular calcium levels (Canossa et al. 1997; Finkbeiner and Greenberg 1997; Jarvis et al. 1997). Finally, postsynaptic properties can be modified by morphological changes such as spine duplication, shrinking or growing, depending on actin modifications (Fischer et al. 1998; Toni et al. 1999). mRNAs encoding β-actin have been localized in neurites and growth cones of cortical cultured neurons (Bassell et al. 1998). This localization of β-actin mRNA and the increase of the amount of protein have been shown to be promoted by NT-3 via a cAMP-dependent mechanism (Zhang et al. 1999). Furthermore, neurotrophins are involved in actin-dependent changes in growth cone shape and motility (Paves and Saarma 1997; Gallo and Letourneau 1998). Hence, neurotrophins might induce rapid and durable modifications of the synaptic transmission by promoting morphological changes in the postsynaptic cell.

Long-lasting or extensive changes in synaptic properties might also result from a synapse-nucleus-synapse shuttle of signals and molecules. Recently, a CREB-based signaling cascade has been shown to mediate the neurotrophin survival signal by the nuclear activation of Bcl-2 (Bonni et al. 1999; Riccio et al. 1999). CREB-mediated, long-term facilitation has been demonstrated in *Aplysia* (Casadio et al. 1999). In hippocampal neurons, mRNAs coding for CREB were detected in dendrites by poly(A) amplification (Crino et al. 1998). Two important facts emerged from this study: first, CREB endogenous protein is present in dendrites but CREB chimeras microperfused in the soma do not translocate in dendrites; second, CREB microperfused in dendrites is transported to the nucleus. Therefore, CREB is most likely to be translated in dendrites.

Neurotrophins can thus exert a modulatory role at the synapse rapidly through local phosphorylations and by promoting actin synthesis and polymerization, on a long time scale via nuclear signaling. Interestingly, mRNAs encoding trk and BDNF have been shown to localize in dendritic fields following synaptic activation (Tongiorgi et al. 1997). CamKII neurotrophins and their receptors can be synthesized in situ. Therefore, these neurotrophic factors can participate in an autocrine loop as well as in a paracrine one, because they can also be secreted and act on presynaptic terminals (Schinder et al. 2000). These findings open up new areas of research that forge hitherto unrecognized links, as between neuronal survival and synaptic plasticity.

6 Conclusions and Perspectives

The signaling pathways allowing synaptic-activity-dependent mRNA transport and local protein synthesis are exciting new research lines. As in other processes, Ca^{2+} seems to be central. The transient modification of Ca^{2+} content in dendrites may activate various different second effectors leading to short-lived and long-lasting changes at the synaptic level. These changes may affect both molecules and structures. The synaptic properties can be modified by changes in the shape of spines resulting from cytoskeletal remodeling, and membrane excitability changes may be modified by insertion of neo-synthesized receptors. Other modifications may also be of importance, such as those exerted on the nucleus by transcription factors synthesized beneath synapses. The synapse, long considered as a purely 'receptive' organelle may have, in the light of these last discoveries, a more complex biology regarding its metabolic autonomy.

The mRNA trafficking and local translation is a mechanism shared by eggs, embryos, plants and neurons. However, little is known about neuron-specificity and/or the messenger-specificity of these mechanisms in neurons. Is mRNA trafficking a phenomenon common to all types of neurons? Are inhibitory and excitatory synapses affected by these mechanisms as well? Are messengers transported in one neuronal type also transported in another? Are messengers translated in the soma, and in dendrites, identical in terms of splicing, editing, or polyadenylation? Studies on RNA binding proteins involved both in splicing and mRNA transport may shed some light on this question. Anatomical and biochemical studies might help to answer questions about the neuronal specificity of the transporting machinery. The combined analysis of different models, different neuronal types and mRNAs, will permit future consolidation of the 'synapse autonomy' model.

References

Ainger K, Avossa D, Morgan F, Hill SJ, Barry C, Barbarese E, Carson JH (1993) Transport and localization of exogenous myelin basic protein mRNA microinjected into oligodendrocytes. J Cell Biol 123:431–441

Albright TD, Jessell TM, Kandel ER, Posner MI (2000) Neural science: a century of progress and the mysteries that remain. Cell 100 [Suppl]:S1–S55

Barbacid M (1994) The Trk family of neurotrophin receptors. J Neurobiol 25:1386–1403

Barbarese E, Koppel DE, Deutscher MP, Smith CL, Ainger K, Morgan F, Carson JH (1995) Protein translation components are colocalized in granules in oligodendrocytes. J Cell Sci 108:2781–2790

Bassell GJ, Zhang H, Byrd AL, Femino AM, Singer RH, Taneja KL, Lifshitz LM, Herman IM, Kosik KS (1998) Sorting of beta-actin mRNA and protein to neurites and growth cones in culture. J Neurosci 18:251–265

Bassell GJ, Oleynikov Y, Singer RH (1999) The travels of mRNAs through all cells large and small. FASEB J 13:447–454

Bennett MK, Erondu NE, Kennedy MB (1983) Purification and characterization of a calmodulin-dependent protein kinase that is highly concentrated in brain. J Biol Chem 258:12735–12744

Bertrand E, Chartrand P, Schaefer M, Shenoy SM, Singer RH, Long RM (1998) Localization of ASH1 mRNA particles in living yeast. Mol Cell 2:437–445

Betz H, Kuhse J, Schmieden V, Laube B, Kirsch J, Harvey RJ (1999) Structure and functions of inhibitory and excitatory glycine receptors. Ann NY Acad Sci 868:667–676

Bilak SR, Morest DK (1998) Differential expression of the metabotropic glutamate receptor mGluR1alpha by neurons and axons in the cochlear nucleus: in situ hybridization and immunohistochemistry. Synapse 28:251–270

Blichenberg A, Schwanke B, Rehbein M, Garner CC, Richter D, Kindler S (1999) Identification of a cis-acting dendritic targeting element in MAP2 mRNAs. J Neurosci 19:8818–8829

Blochl A, Thoenen H (1996) Localization of cellular storage compartments and sites of constitutive and activity-dependent release of nerve growth factor (NGF) in primary cultures of hippocampal neurons. Mol Cell Neurosci 7:173–190

Bodian D (1965) A suggestive relationship of nerve cell RNA with specific synaptic sites. Proc Natl Acad Sci USA 76:5982–5986

Bole DG, Hendershot LM, Kearney JF (1986) Posttranslational association of immunoglobulin heavy chain binding protein with nascent heavy chains in nonsecreting and secreting hybridomas. J Cell Biol 102:1558–1566

Bonni A, Brunet A, West AE, Datta SR, Takasu MA, Greenberg ME (1999) Cell survival promoted by the Ras-MAPK signaling pathway by transcription-dependent and -independent mechanisms. Science 286:1358–1362

Bowlby MR, Fadool DA, Holmes TC, Levitan IB (1997) Modulation of the Kv1.3 potassium channel by receptor tyrosine kinases. J Gen Physiol 110:601–610

Brosius J (1999) RNAs from all categories generate retrosequences that may be exapted as novel genes or regulatory elements. Gene 238:115–134

Bruckenstein DA, Lein PJ, Higgins D, Fremeau RT Jr (1990) Distinct spatial localization of specific mRNAs in cultured sympathetic neurons. Neuron 5:809–819

Buckanovich RJ, Posner JB, Darnell RB (1993) Nova, the paraneoplastic Ri antigen, is homologous to an RNA-binding protein and is specifically expressed in the developing motor system. Neuron 11:657–672

Burd CG, Dreyfuss G (1994) Conserved structures and diversity of functions of RNA-binding proteins. Science 265:615–621

Burgin KE, Waxham MN, Rickling S, Westgate SA, Mobley WC, Kelly PT (1990) In situ hybridization histochemistry of Ca^{2+}/calmodulin-dependent protein kinase in developing rat brain. J Neurosci 10:1788–1798

Canossa M, Griesbeck O, Berninger B, Campana G, Kolbeck R, Thoenen H (1997) Neurotrophin release by neurotrophins: implications for activity- dependent neuronal plasticity. Proc Natl Acad Sci USA 94:13279–13286

Casadio A, Martin KC, Giustetto M, Zhu H, Chen M, Bartsch D, Bailey CH, Kandel ER (1999) A transient, neuron-wide form of CREB-mediated long-term facilitation can be stabilized at specific synapses by local protein synthesis. Cell 99:221–237

Chicurel ME, Harris KM (1992) Three-dimensional analysis of the structure and composition of CA3 branched dendritic spines and their synaptic relationships with mossy fiber boutons in the rat hippocampus. J Comp Neurol 325:169–182

Chicurel ME, Terrian DM, Potter H (1993) mRNA at the synapse: analysis of a synaptosomal preparation enriched in hippocampal dendritic spines. J Neurosci 13:4054–4063

Crino P, Khodakhah K, Becker K, Ginsberg S, Hemby S, Eberwine J (1998) Presence and phosphorylation of transcription factors in developing dendrites. Proc Natl Acad Sci USA 95:2313–2318

Darnell RB (1996) Onconeural antigens and the paraneoplastic neurologic disorders: at the intersection of cancer, immunity, and the brain. Proc Natl Acad Sci USA 93:4529–4536

Davis L, Banker GA, Steward O (1987) Selective dendritic transport of RNA in hippocampal neurons in culture. Nature 330:477–479

De Boulle K, Verkerk AJ, Reyniers E, Vits L, Hendrickx J, Van Roy B, Van den Bos F, de Graaff E, Oostra BA, Willems PJ (1993) A point mutation in the FMR-1 gene associated with fragile X mental retardation. Nat Genet 3:31–35

De Camilli P, Moretti M, Donini SD, Walter U, Lohmann SM (1986) Heterogeneous distribution of the cAMP receptor protein RII in the nervous system: evidence for its intracellular accumulation on microtubules, microtubule-organizing centers, and in the area of the Golgi complex. J Cell Biol 103:189–203

DeChiara TM, Brosius J (1987) Neural BC1 RNA: cDNA clones reveal nonrepetitive sequence content. Proc Natl Acad Sci USA 84:2624–2628

Decker CJ, Parker R (1995) Diversity of cytoplasmic functions for the 3′ untranslated region of eukaryotic transcripts. Curr Opin Cell Biol 7:386–392

Deshler JO, Highett MI, Schnapp BJ (1997) Localization of *Xenopus* Vg1 mRNA by Vera protein and the endoplasmic reticulum. Science 276:1128–1131

Dotti CG, Simons K (1990) Polarized sorting of viral glycoproteins to the axon and dendrites of hippocampal neurons in culture. Cell 62:63–72

Elkon KB, Parnassa AP, Foster CL (1985) Lupus autoantibodies target ribosomal P proteins. J Exp Med 162:459–471

Etkin LD, Lipshitz HD (1999) RNA localization. FASEB J 13:419–420

Feng Y, Absher D, Eberhart DE, Brown V, Malter HE, Warren ST (1997a) FMRP associates with polyribosomes as an mRNP, and the I304 N mutation of severe fragile X syndrome abolishes this association. Mol Cell 1:109–118

Feng Y, Gutekunst CA, Eberhart DE, Yi H, Warren ST, Hersch SM (1997b) Fragile X mental retardation protein: nucleocytoplasmic shuttling and association with somatodendritic ribosomes. J Neurosci 17:1539–1547

Ferrandon D, Elphick L, Nusslein-Volhard C, St Johnston D (1994) Staufen protein associates with the 3′UTR of bicoid mRNA to form particles that move in a microtubule-dependent manner. Cell 79:1221–1232

Ferrandon D, Koch I, Westhof E, Nusslein-Volhard C (1997) RNA-RNA interaction is required for the formation of specific bicoid mRNA 3' UTR-STAUFEN ribonucleoprotein particles. EMBO J 16:1751-1758

Finkbeiner S, Greenberg ME (1997) Spatial features of calcium-regulated gene expression. Bioessays 19:657-660

Fischer M, Kaech S, Knutti D, Matus A (1998) Rapid actin-based plasticity in dendritic spines. Neuron 20:847-854

Furuichi T, Simon-Chazottes D, Fujino I, Yamada N, Hasegawa M, Miyawaki A, Yoshikawa S, Guenet JL, Mikoshiba K (1993) Widespread expression of inositol 1,4,5-trisphosphate receptor type 1 gene (Insp3r1) in the mouse central nervous system. Recept Channels 1:11-24

Gallo G, Letourneau PC (1998) Localized sources of neurotrophins initiate axon collateral sprouting. J Neurosci 18:5403-5414

Gardiol A, Racca C, Triller A (1999) Dendritic and postsynaptic protein synthetic machinery. J Neurosci 19:168-179

Garner CC, Tucker RP, Matus A (1988) Selective localization of messenger RNA for cytoskeletal protein MAP2 in dendrites. Nature 336:674-677

Gazzaley AH, Benson DL, Huntley GW, Morrison JH (1997) Differential subcellular regulation of NMDAR1 protein and mRNA in dendrites of dentate gyrus granule cells after perforant path transection. J Neurosci 17:2006-2017

Green WN, Millar NS (1995) Ion-channel assembly. Trends Neurosci 18:280-287

Hoch W, Betz H, Becker CM (1989) Primary cultures of mouse spinal cord express the neonatal isoform of the inhibitory glycine receptor. Neuron 3:339-348

Jansen RP (1999) RNA-cytoskeletal associations. FASEB J 13:455-466

Jarvis CR, Xiong ZG, Plant JR, Churchill D, Lu WY, MacVicar BA, MacDonald JF (1997) Neurotrophin modulation of NMDA receptors in cultured murine and isolated rat neurons. J Neurophysiol 78:2363-2371

Jasmin BJ, Cartaud J, Bornens M, Changeux JP (1989) Golgi apparatus in chick skeletal muscle: changes in its distribution during end plate development and after denervation. Proc Natl Acad Sci USA 86:7218-7222

Jasmin BJ, Antony C, Changeux JP, Cartaud J (1995) Nerve-dependent plasticity of the Golgi complex in skeletal muscle fibres: compartmentalization within the subneural sarcoplasm. Eur J Neurosci 7:470-479

Kaiserman-Abramof IR, Palay SL (1969) Fine structural studies of the cerebellar cotex in a normyrid fish. In: Llinas R (ed) Neurobiology of cerebellar evolution and development. American Medical Association, Chicago, pp 171-205

Kang H, Schuman EM (1996) A requirement for local protein synthesis in neurotrophin-induced hippocampal synaptic plasticity. Science 273:1402-1406

Kelly PT, McGuinness TL, Greengard P (1984) Evidence that the major postsynaptic density protein is a component of a Ca^{2+}/calmodulin-dependent protein kinase. Proc Natl Acad Sci USA 81:945-949

Kiebler MA, DesGroseillers L (2000) Molecular insights into mRNA transport and local translation in the mammalian nervous system. Neuron 25:19-28

Kiebler MA, Hemraj I, Verkade P, Kohrmann M, Fortes P, Marion RM, Ortin J, Dotti CG (1999) The mammalian staufen protein localizes to the somatodendritic domain of cultured hippocampal neurons: implications for its involvement in mRNA transport. J Neurosci 19:288-297

Kindler S, Muller R, Chung WJ, Garner CC (1996) Molecular characterization of dendritically localized transcripts encoding MAP2. Brain Res Mol Brain Res 36:63-69

Kleiman R, Banker G, Steward O (1990) Differential subcellular localization of particular mRNAs in hippocampal neurons in culture. Neuron 5:821-830

Kleiman R, Banker G, Steward O (1993a) Inhibition of protein synthesis alters the subcellular distribution of mRNA in neurons but does not prevent dendritic transport of RNA. Proc Natl Acad Sci USA 90:11192-11196

Kleiman R, Banker G, Steward O (1993b) Subcellular distribution of rRNA and poly(A) RNA in hippocampal neurons in culture. Brain Res Mol Brain Res 20:305–312

Kleiman R, Banker G, Steward O (1994) Development of subcellular mRNA compartmentation in hippocampal neurons in culture. J Neurosci 14:1130–1140

Knowles RB, Sabry JH, Martone ME, Deerinck TJ, Ellisman MH, Bassell GJ, Kosik KS (1996) Translocation of RNA granules in living neurons. J Neurosci 16:7812–7820

Kohrmann M, Luo M, Kaether C, DesGroseillers L, Dotti CG, Kiebler MA (1999) Microtubule-dependent recruitment of Staufen-green fluorescent protein into large RNA-containing granules and subsequent dendritic transport in living hippocampal neurons. Mol Biol Cell 10:2945–2953

Korkotian E, Segal M (1999) Release of calcium from stores alters the morphology of dendritic spines in cultured hippocampal neurons. Proc Natl Acad Sci USA 96:12068–12072

Lasko P (1999) RNA sorting in *Drosophila* oocytes and embryos. FASEB J 13:421–433

Lawrence JB, Singer RH (1986) Intracellular localization of messenger RNAs for cytoskeletal proteins. Cell 45:407–415

Legendre P (1997) Pharmacological evidence for two types of postsynaptic glycinergic receptors on the Mauthner cell of 52-h-old zebrafish larvae. J Neurophysiol 77:2400–2415

Lesser SS, Lo DC (1995) Regulation of voltage-gated ion channels by NGF and ciliary neurotrophic factor in SK-N-SH neuroblastoma cells. J Neurosci 15:253–261

Link W, Konietzko U, Kauselmann G, Krug M, Schwanke B, Frey U, Kuhl D (1995) Somatodendritic expression of an immediate early gene is regulated by synaptic activity. Proc Natl Acad Sci USA 92:5734–5738

Llano I, DiPolo R, Marty A (1994) Calcium-induced calcium release in cerebellar Purkinje cells. Neuron 12:663–673

Lledo PM, Zhang X, Sudhof TC, Malenka RC, Nicoll RA (1998) Postsynaptic membrane fusion and long-term potentiation. Science 279:399–403

Lowenstein PR, Morrison EE, Bain D, Shering AF, Banting G, Douglas P, Castro MG (1994) Polarized distribution of the trans-Golgi network marker TGN38 during the in vitro development of neocortical neurons: effects of nocodazole and brefeldin A. Eur J Neurosci 6:1453–1465

Lu Z, McLaren RS, Winters CA, Ralston E (1998) Ribosome association contributes to restricting mRNAs to the cell body of hippocampal neurons. Mol Cell Neurosci 12:363–375

Luscher C, Xia H, Beattie EC, Carroll RC, von Zastrow M, Malenka RC, Nicoll RA (1999) Role of AMPA receptor cycling in synaptic transmission and plasticity. Neuron 24:649–658

Lyford GL, Yamagata K, Kaufmann WE, Barnes CA, Sanders LK, Copeland NG, Gilbert DJ, Jenkins NA, Lanahan AA, Worley PF (1995) Arc, a growth factor and activity-regulated gene, encodes a novel cytoskeleton-associated protein that is enriched in neuronal dendrites. Neuron 14:433–445

Maletic-Savatic M, Malinow R (1998) Calcium-evoked dendritic exocytosis in cultured hippocampal neurons, part I. *trans*-Golgi network-derived organelles undergo regulated exocytosis. J Neurosci 18:6803–6813

Maletic-Savatic M, Koothan T, Malinow R (1998) Calcium-evoked dendritic exocytosis in cultured hippocampal neurons, part II. Mediation by calcium/calmodulin-dependent protein kinase II. J Neurosci 18:6814–6821

Marin P, Nastiuk KL, Daniel N, Girault JA, Czernik AJ, Glowinski J, Nairn AC, Premont J (1997) Glutamate-dependent phosphorylation of elongation factor-2 and inhibition of protein synthesis in neurons. J Neurosci 17:3445–3454

Marion RM, Fortes P, Beloso A, Dotti C, Ortin J (1999) A human sequence homologue of Staufen is an RNA-binding protein that is associated with polysomes and localizes to the rough endoplasmic reticulum. Mol Cell Biol 19:2212–2219

Martin KC, Casadio A, Zhu H, E Y, Rose JC, Chen M, Bailey CH, Kandel ER (1997) Synapse-specific, long-term facilitation of *Aplysia* sensory to motor synapses: a function for local protein synthesis in memory storage. Cell 91:927–938

Martone ME, Pollock JA, Jones YZ, Ellisman MH (1996) Ultrastructural localization of dendritic messenger RNA in adult rat hippocampus. J Neurosci 16:7437–7446

Mathisen PM, Johnson JM, Kawczak JA, Tuohy VK (1999) Visinin-like protein (VILIP) is a neuron-specific calcium-dependent double-stranded RNA-binding protein. J Biol Chem 274:31571–31576

Mayford M, Bach ME, Huang YY, Wang L, Hawkins RD, Kandel ER (1996a) Control of memory formation through regulated expression of a CaMKII transgene. Science 274:1678–1683

Mayford M, Baranes D, Podsypanina K, Kandel ER (1996b) The 3'-untranslated region of CaMKII alpha is a cis-acting signal for the localization and translation of mRNA in dendrites. Proc Natl Acad Sci USA 93:13250–13255

McAllister AK, Katz LC, Lo DC (1999) Neurotrophins and synaptic plasticity. Annu Rev Neurosci 22:295–318

Meyer G, Kirsch J, Betz H, Langosch D (1995) Identification of a gephyrin binding motif on the glycine receptor beta subunit. Neuron 15:563–572

Miyashiro K, Dichter M, Eberwine J (1994) On the nature and differential distribution of mRNAs in hippocampal neurites: implications for neuronal functioning. Proc Natl Acad Sci USA 91:10800–10804

Mohr E (1999) Subcellular RNA compartmentalization. Prog Neurobiol 57:507–525

Mowry KL, Cote CA (1999) RNA sorting in Xenopus oocytes and embryos. FASEB J 13:435–445

Muslimov IA, Banker G, Brosius J, Tiedge H (1998) Activity-dependent regulation of dendritic BC1 RNA in hippocampal neurons in culture. J Cell Biol 141:1601–1611

Muslimov IA, Santi E, Homel P, Perini S, Higgins D, Tiedge H (1997) RNA transport in dendrites: a cis-acting targeting element is contained within neuronal BC1 RNA. J Neurosci 17:4722–4733

Nusser Z, Lujan R, Laube G, Roberts JD, Molnar E, Somogyi P (1998) Cell type and pathway dependence of synaptic AMPA receptor number and variability in the hippocampus. Neuron 21:545–559

Ouyang Y, Rosenstein A, Kreiman G, Schuman EM, Kennedy MB (1999) Tetanic stimulation leads to increased accumulation of Ca(2+)/calmodulin-dependent protein kinase II via dendritic protein synthesis in hippocampal neurons. J Neurosci 19:7823–7833

Paves H, Saarma M (1997) Neurotrophins as in vitro growth cone guidance molecules for embryonic sensory neurons. Cell Tissue Res 290:285–297

Pelham HR, Munro S (1993) Sorting of membrane proteins in the secretory pathway. Cell 75:603–605

Peters A, Palay SL, Webster HdF (1991) The fine structure of the nervous system. Neurons and their supporting cells, 3rd edn. Oxford Univ Press, Oxford

Pierce JP, van Leyen K, McCarthy JB (2000) Translocation machinery for synthesis of integral membrane and secretory proteins in dendritic spines. Nat Neurosci 3:311–313

Posner JB, Dalmau JO (1997) Paraneoplastic syndromes affecting the central nervous system. Annu Rev Med 48:157–166

Pozzo-Miller LD, Pivovarova NB, Leapman RD, Buchanan RA, Reese TS, Andrews SB (1997) Activity-dependent calcium sequestration in dendrites of hippocampal neurons in brain slices. J Neurosci 17:8729–8738

Prakash N, Fehr S, Mohr E, Richter D (1997) Dendritic localization of rat vasopressin mRNA: ultrastructural analysis and mapping of targeting elements. Eur J Neurosci 9:523–532

Prostko CR, Dholakia JN, Brostrom MA, Brostrom CO (1995) Activation of the double-stranded RNA-regulated protein kinase by depletion of endoplasmic reticular calcium stores. J Biol Chem 270:6211–6215

Quinlan EM, Philpot BD, Huganir RL, Bear MF (1999) Rapid, experience-dependent expression of synaptic NMDA receptors in visual cortex in vivo. Nat Neurosci 2:352–357

Racca C, Gardiol A, Triller A (1997a) Dendritic and postsynaptic localizations of glycine receptor alpha subunit mRNAs. J Neurosci 17:1691–1700

Racca C, Gardiol A, Triller A (1997b) Nova-1: an RNA binding protein associated with GlyR α2 subunit mRNA in rat spinal cord. Soc Neurosci Abstr 23:486.4

Racca C, Stephenson FA, Streit P, Roberts JD, Somogyi P (2000) NMDA receptor content of synapses in stratum radiatum of the hippocampal CA1 area. J Neurosci 20:2512–2522

Rao A, Steward O (1991) Evidence that protein constituents of postsynaptic membrane specializations are locally synthesized: analysis of proteins synthesized within synaptosomes. J Neurosci 11:2881–2895

Rhoads RE (1993) Regulation of eukaryotic protein synthesis by initiation factors. J Biol Chem 268:3017–3020

Riccio A, Ahn S, Davenport CM, Blendy JA, Ginty DD (1999) Mediation by a CREB family transcription factor of NGF-dependent survival of sympathetic neurons. Science 286:2358–2361

Rongo C, Kaplan JM (1999) CaMKII regulates the density of central glutamatergic synapses in vivo. Nature 402:195–199

Rosenbluth J (1962) Subsurface cisterns and their relationship to the neuronal plasma membrane. J Cell Biol 13:405–421

Rossier MF, Putney JW Jr (1991) The identity of the calcium-storing, inositol 1,4,5-trisphosphate-sensitive organelle in non-muscle cells: calciosome, endoplasmic reticulum ... or both? Trends Neurosci 14:310–314

Rubio ME, Wenthold RJ (1999) Differential distribution of intracellular glutamate receptors in dendrites. J Neurosci 19:5549–5562

Ryazanov AG (1987) Ca^{2+}/calmodulin-dependent phosphorylation of elongation factor 2. FEBS Lett 214:331–334

Ryazanov AG, Shestakova EA, Natapov PG (1988) Phosphorylation of elongation factor 2 by EF-2 kinase affects rate of translation. Nature 334:170–173

Sanes JR, Lichtman JW (1999) Development of the vertebrate neuromuscular junction. Annu Rev Neurosci 22:389–442

Saraste J, Lahtinen U, Goud B (1995) Localization of the small GTP-binding protein rab1p to early compartments of the secretory pathway. J Cell Sci 108:1541–1552

Schacher S, Glanzman D, Barzilai A, Dash P, Grant SG, Keller F, Mayford M, Kandel ER (1990) Long-term facilitation in *Aplysia*: persistent phosphorylation and structural changes. Cold Spring Harbor Symp Quant Biol 55:187–202

Scheetz AJ, Nairn AC, Constantine-Paton M (1997) N-methyl-D-aspartate receptor activation and visual activity induce elongation factor-2 phosphorylation in amphibian tecta: a role for N- methyl-D-aspartate receptors in controlling protein synthesis. Proc Natl Acad Sci USA 94:14770–14775

Scheetz AJ, Nairn AC, Constantine-Paton M (2000) NMDA receptor-mediated control of protein synthesis at developing synapses. Nat Neurosci 3:211–216

Schinder AF, Berninger B, Poo M (2000) Postsynaptic target specificity of neurotrophin-induced presynaptic potentiation. Neuron 25:151–163

Schmieden V, Grenningloh G, Schofield PR, Betz H (1989) Functional expression in *Xenopus* oocytes of the strychnine binding 48 kd subunit of the glycine receptor. EMBO J 8:695–700

Schorge S, Gupta S, Lin Z, McEnery MW, Lipscombe D (1999) Calcium channel activation stabilizes a neuronal calcium channel mRNA. Nat Neurosci 2:785–790

Segal I, Korkotian I, Murphy DD (2000) Dendritic spine formation and pruning: common cellular mechanisms? Trends Neurosci 23:53–57

Sharma N, D'Arcangelo G, Kleinlaus A, Halegoua S, Trimmer JS (1993) Nerve growth factor regulates the abundance and distribution of K^+ channels in PC12 cells. J Cell Biol 123:1835–1843

Shi SH, Hayashi Y, Petralia RS, Zaman SH, Wenthold RJ, Svoboda K, Malinow R (1999) Rapid spine delivery and redistribution of AMPA receptors after synaptic NMDA receptor activation. Science 284:1811–1816

Silva AJ, Paylor R, Wehner JM, Tonegawa S (1992a) Impaired spatial learning in alpha-calcium-calmodulin kinase II mutant mice. Science 257:206–211

Silva AJ, Stevens CF, Tonegawa S, Wang Y (1992b) Deficient hippocampal long-term potentiation in alpha-calcium- calmodulin kinase II mutant mice. Science 257:201–206

Singer RH, Langevin GL, Lawrence JB (1989) Ultrastructural visualization of cytoskeletal mRNAs and their associated proteins using double-label in situ hybridization. J Cell Biol 108:2343–2353

Siomi H, Siomi MC, Nussbaum RL, Dreyfuss G (1993) The protein product of the fragile X gene, FMR1, has characteristics of an RNA-binding protein. Cell 74:291–298

Siomi H, Choi M, Siomi MC, Nussbaum RL, Dreyfuss G (1994) Essential role for KH domains in RNA binding: impaired RNA binding by a mutation in the KH domain of FMR1 that causes fragile X syndrome. Cell 77:33–39

Skryabin BV, Kremerskothen J, Vassilacopoulou D, Disotell TR, Kapitonov VV, Jurka J, Brosius J (1998) The BC200 RNA gene and its neural expression are conserved in Anthropoidea (Primates). J Mol Evol 47:677–685

Spacek J, Harris KM (1997) Three-dimensional organization of smooth endoplasmic reticulum in hippocampal CA1 dendrites and dendritic spines of the immature and mature rat. J Neurosci 17:190–203

Srivastava SP, Davies MV, Kaufman RJ (1995) Calcium depletion from the endoplasmic reticulum activates the double- stranded RNA-dependent protein kinase (PKR) to inhibit protein synthesis. J Biol Chem 270:16619–16624

St Johnston D, Nusslein-Volhard C (1992) The origin of pattern and polarity in the *Drosophila* embryo. Cell 68:201–219

Steward O (1987) Regulation of synaptogenesis through the local synthesis of protein at the postsynaptic site. Prog Brain Res 71:267–279

Steward O (1997) mRNA localization in neurons: a multipurpose mechanism? Neuron 18:9–12

Steward O, Falk PM (1985) Polyribosomes under developing spine synapses: growth specializations of dendrites at sites of synaptogenesis. J Neurosci Res 13:75–88

Steward O, Falk PM (1986) Protein-synthetic machinery at postsynaptic sites during synaptogenesis: a quantitative study of the association between polyribosomes and developing synapses. J Neurosci 6:412–423

Steward O, Halpain S (1999) Lamina-specific synaptic activation causes domain-specific alterations in dendritic immunostaining for MAP2 and CAM kinase II. J Neurosci 19:7834–7845

Steward O, Levy WB (1982) Preferential localization of polyribosomes under the base of dendritic spines in granule cells of the dentate gyrus. J Neurosci 2:284–291

Steward O, Bakker CE, Willems PJ, Oostra BA (1998a) No evidence for disruption of normal patterns of mRNA localization in dendrites or dendritic transport of recently synthesized mRNA in FMR1 knockout mice, a model for human fragile-X mental retardation syndrome. Neuroreport 9:477–481

Steward O, Wallace CS, Lyford GL, Worley PF (1998b) Synaptic activation causes the mRNA for the IEG Arc to localize selectively near activated postsynaptic sites on dendrites. Neuron 21:741–751

Stowell JN, Craig AM (1999) Axon/dendrite targeting of metabotropic glutamate receptors by their cytoplasmic carboxy-terminal domains. Neuron 22:525–536

Suen PC, Wu K, Levine ES, Mount HT, Xu JL, Lin SY, Black IB (1997) Brain-derived neurotrophic factor rapidly enhances phosphorylation of the postsynaptic N-methyl-D-aspartate receptor subunit 1. Proc Natl Acad Sci USA 94:8191–8195

Sundell CL, Singer RH (1991) Requirement of microfilaments in sorting of actin messenger RNA. Science 253:1275–1277

Takahashi T, Momiyama A, Hirai K, Hishinuma F, Akagi H (1992) Functional correlation of fetal and adult forms of glycine receptors with developmental changes in inhibitory synaptic receptor channels. Neuron 9:1155–1161

Takei K, Stukenbrok H, Metcalf A, Mignery GA, Sudhof TC, Volpe P, De Camilli P (1992) Ca²⁺ stores in Purkinje neurons: endoplasmic reticulum subcompartments demonstrated by the heterogeneous distribution of the InsP3 receptor, Ca(2+)-ATPase, and calsequestrin. J Neurosci 12:489–505

Tiedge H, Brosius J (1996) Translational machinery in dendrites of hippocampal neurons in culture. J Neurosci 16:7171–7181

Tiedge H, Fremeau RT Jr, Weinstock PH, Arancio O, Brosius J (1991) Dendritic location of neural BC1 RNA. Proc Natl Acad Sci USA 88:2093–2097

Tiedge H, Zhou A, Thorn NA, Brosius J (1993) Transport of BC1 RNA in hypothalamo-neurohypophyseal axons. J Neurosci 13:4214–4219

Timpe LC, Fantl WJ (1994) Modulation of a voltage-activated potassium channel by peptide growth factor receptors. J Neurosci 14:1195–1201

Tongiorgi E, Righi M, Cattaneo A (1997) Activity-dependent dendritic targeting of BDNF and TrkB mRNAs in hippocampal neurons. J Neurosci 17:9492–9505

Toni N, Buchs PA, Nikonenko I, Bron CR, Muller D (1999) LTP promotes formation of multiple spine synapses between a single axon terminal and a dendrite. Nature 402:421–425

Torre ER, Steward O (1992) Demonstration of local protein synthesis within dendrites using a new cell culture system that permits the isolation of living axons and dendrites from their cell bodies. J Neurosci 12:762–772

Torre ER, Steward O (1996) Protein synthesis within dendrites: glycosylation of newly synthesized proteins in dendrites of hippocampal neurons in culture. J Neurosci 16:5967–5978

Triller A, Cluzeaud F, Pfeiffer F, Betz H, Korn H (1985) Distribution of glycine receptors at central synapses: an immunoelectron microscopy study. J Cell Biol 101:683–688

Villa A, Podini P, Clegg DO, Pozzan T, Meldolesi J (1991) Intracellular Ca^{2+} stores in chicken Purkinje neurons: differential distribution of the low affinity-high capacity Ca^{2+} binding protein, calsequestrin, of Ca^{2+} ATPase and of the ER lumenal protein, Bip. J Cell Biol 113:779–791

Wallace CS, Lyford GL, Worley PF, Steward O (1998) Differential intracellular sorting of immediate early gene mRNAs depends on signals in the mRNA sequence. J Neurosci 18:26–35

Weiler IJ, Greenough WT (1993) Metabotropic glutamate receptors trigger postsynaptic protein synthesis. Proc Natl Acad Sci USA 90:7168–7171

Weiler IJ, Childers WS, Greenough WT (1996) Calcium ion impedes translation initiation at the synapse. J Neurochem 66:197–202

Weiler IJ, Irwin SA, Klintsova AY, Spencer CM, Brazelton AD, Miyashiro K, Comery TA, Patel B, Eberwine J, Greenough WT (1997) Fragile X mental retardation protein is translated near synapses in response to neurotransmitter activation. Proc Natl Acad Sci USA 94:5395–5400

Wickham L, Duchaine T, Luo M, Nabi IR, DesGroseillers L (1999) Mammalian staufen is a double-stranded-RNA- and tubulin-binding protein which localizes to the rough endoplasmic reticulum. Mol Cell Biol 19:2220–2230

Wilde A, Reaves B, Banting G (1992) Epitope mapping of two isoforms of a trans Golgi network specific integral membrane protein TGN38/41. FEBS Lett 313:235–238

Wilhelm JE, Vale RD (1993) RNA on the move: the mRNA localization pathway. J Cell Biol 123:269–274

Wu L, Wells D, Tay J, Mendis D, Abbott MA, Barnitt A, Quinlan E, Heynen A, Fallon JR, Richter JD (1998) CPEB-mediated cytoplasmic polyadenylation and the regulation of experience-dependent translation of alpha-CaMKII mRNA at synapses. Neuron 21:1129–1139

Zamanillo D, Sprengel R, Hvalby O, Jensen V, Burnashev N, Rozov A, Kaiser KM, Koster HJ, Borchardt T, Worley P, Lubke J, Frotscher M, Kelly PH, Sommer B, Andersen P, Seeburg PH, Sakmann B (1999) Importance of AMPA receptors for hippocampal synaptic plasticity but not for spatial learning. Science 284:1805–1811

Zhang HL, Singer RH, Bassell GJ (1999) Neurotrophin regulation of beta-actin mRNA and protein localization within growth cones. J Cell Biol 147:59–70

Neuronal BC1 RNA: Intracellular Transport and Activity-Dependent Modulation

Jürgen Brosius[1] and Henri Tiedge[2]

RNA transport in dendrites, and subsequent local translation in postsynaptic microdomains, have increasingly been recognized as potentially powerful tools for neurons in long-term modulations of individual synapses (see reviews by Steward 1997; Kuhl and Skehel 1998; Tiedge et al. 1999). While analogous mechanisms have also been reported to be employed by other cell types (reviewed by Bassell et al. 1999), neurons face a particular challenge in that thousands of synapses per dendritic arborization (many of which at considerable distances from the soma) have to be regulated in an independent and input-specific manner. It now appears plausible that such regulation is achieved by two mechanisms: (1) targeted delivery of specific proteins to dendritic sites of demand, and (2) dendritic transport of select RNAs, followed by local postsynaptic translation when and where needed.

The concept of local protein synthesis in dendrites has been strengthened by the identification of a number of mRNAs that are selectively targeted to dendrites. Dendritic mRNAs encode proteins that belong to diverse classes, including cytoskeletal components, kinases, and receptors (for examples, see reviews by Steward 1997; Kuhl and Skehel 1998; Tiedge et al. 1999). Components of the protein synthetic machinery, including initiation and elongation factors as well as factors required in cotranslational protein sorting mechanisms, have been detected in dendrites of various types of neurons (Tiedge and Brosius 1996; Gardiol et al. 1999). The capacity for local protein synthesis has been demonstrated in dendrites in vitro (Torre and Steward 1992) and in vivo (Mayford et al. 1996), as has the potential for dendritic glycosylation of locally synthesized proteins (Torre and Steward 1996).

There is thus solid evidence for translational machinery being operational in dendrites. However, as dendritic translation, and translation-related processes, are likely to be subject to activity-dependent regulation, it can be anticipated that aspects of the translational apparatus in dendrites be neuron- or even dendrite-specific. We have previously identified neuronal BC1 RNA, a short untranslatable RNA (DeChiara and Brosius 1987) that is specifically

[1] Institute of Experimental Pathology/Molecular Neurobiology, Center for Molecular Biology of Inflammation, University of Münster, 48149 Münster, Germany
[2] Department of Physiology and Pharmacology, Department of Neurology, State University of New York, Health Science Center at Brooklyn, Brooklyn, New York 11203, USA

Results and Problems in Cell Differentiation, Vol. 34
D. Richter (Ed.): Cell Polarity and Subcellular RNA Localization
© Springer-Verlag Berlin Heidelberg 2001

localized to dendrites (Tiedge et al. 1991, 1992). BC1 RNA is complexed with proteins to form a RNPs (ribonucleoprotein particle; Kobayashi et al. 1991; Cheng et al. 1996), and it has been suggested to play a functional role in regulation of translation or translation-related processes in postsynaptic microdomains (Brosius and Tiedge 1995). In the following, we will summarize experimental evidence from the last few years that has advanced our under-standing of evolutionary relationships, intracellular transport, and activity-dependent regulation of neuronal BC1 RNA.

1 Evolutionary Relationships

Neuronal BC1 RNA has been generated by reverse transcription of a tRNA^{Ala} and random integration of its cDNA copy (reviewed by Brosius 1999a,b) in a mechanism termed retroposition (Weiner et al. 1986). The gene arose after the mammalian radiation but prior to the diversification of the order Rodentia (Martignetti and Brosius 1993) between 110 and 60 million years ago. The locus of integration corresponds to distal chromosome 7 near *Fgf3* in the mouse (Taylor et al. 1997) where flanking promoter elements were located or sub-sequently created. In conjunction with the internal promoter elements (boxes A + B'; Martignetti and Brosius 1995), this allowed efficient and specific transcription of BC1 RNA in neurons. In most other locations the new tRNA retronuon (a nuon is any definable nucleic acid sequence; see Brosius and Gould 1992, 1993) would have been transcriptionally silent – a fate it would have shared with most retrosequences including SINEs (short interspersed repetitive elements; see Brosius 1999a,b). In the meantime, the sequence of BC1 RNA changed and existing proteins were recruited from the cell to bind BC1 RNA, forming an RNP. This RNP was exapted (recruited; Gould and Vrba 1982) into a function. Although BC1 RNA is restricted to rodents, the coding region of BC1 RNA is far more conserved among different rodent species than the respective 5' or 3' flanking sequences (Martignetti and Brosius 1993). This strongly supports the notion that the gene product has a function, and that the gene is under selective pressure.

The BC1 RNA transcript comprises about 152 nucleotides, twice as many as tRNA^{Ala}. The sequence similarity with tRNA^{Ala}, limited to the 5' half of BC1 RNA, is about 80%. However, this domain does not fold in a tRNA-like manner, but rather forms a stable stem/loop. The 3' half of BC1 RNA consists of a central A-rich region and a unique region. The unique region derived its name from the fact that, as a probe in DNA blots at medium stringency, it identifies only a single band; no other sequences resembling the unique region have been found in rodent genomes. This is consistent with the fact that there is a single gene encoding BC1 RNA in most rodents examined. In contrast, A-rich regions are frequently found in genomes of higher Eukarya. The A-rich region may stem from aberrant polyadenylation of the tRNA^{Ala} template (see, for example, Lu and Bablanian 1996). Alternatively, it may have been recruited – along with the unique region – from the integration locus, a notion consistent with the fact that transcription proceeds beyond the tRNA domain until it encounters a stretch of T-residues that serves as termination signal for RNA polymerase III.

We have identified BC1 RNA as the first master gene of a SINEs. A product of a single gene, BC1 RNA was thus reverse-transcribed and integrated into the genome in tens of thousands of separate events, leading to the dispersion of a subfamily of ID repetitive elements (DeChiara and Brosius 1987; Deininger et al. 1996). Due to the lack of upstream promoter elements in their new integration loci, virtually all of these ID elements are transcriptionally silent. Nevertheless, ID elements may modulate expression of targeted genes (Vidal and Cuzin 1995). Not infrequently, SINEs such as Alu, B1, and B2 elements can influence diverse processes in gene expression, including enhancement or silencing of transcription, modulation of mRNA stability, and providing (alternative) splice sites, polyadenylation signals and/or novel small protein coding domains (Brosius 1999a,b).

2 Intracellular Transport

BC1 RNA was initially detected, using in situ hybridization, in dendritic layers in brain and in dendrites of acutely isolated spinal cord neurons (Tiedge et al. 1991, 1992). These observations were later extended to sympathetic and hippocampal neurons in primary culture (unpubl. data and Muslimov et al. 1998). BC1 RNA was also detected in preparations of synaptodendrosomes where it was found to be co-enriched, at substantial levels, with various dendritic mRNAs (Chicurel et al. 1993; Rao and Steward 1993). The combined results thus raised the question as to what mechanism selectively targets BC1 RNA to dendrites.

To address this question, Muslimov et al. (1997) used a microinjection paradigm with sympathetic neurons in culture, analogous to the approach by Ainger et al. (1993, 1997) with cultured oligodendrocytes. BC1 RNA and chimeric RNAs containing BC1 RNA, microinjected into the perinuclear cytoplasmic region of cultured sympathetic neurons, were specifically and rapidly delivered to dendrites (Fig. 1). Injected nuclear U4 RNA, or random sequence RNAs of similar length, did not exit somata to any significant extent. The targeted transport of BC1 RNA to dendrites was rapid: the calculated initial velocity was about 400 μm/h. This transport rate, and recent observations that the dendritic delivery of BC1 RNA is dependent upon intact microtubules (unpubl.), indicate that a motor molecule of the kinesin superfamily is mediating dendritic BC1 transport.

Subsequent experiments showed that the dendritic targeting competence of BC1 RNA resides within its 5' domain (Fig. 2; Muslimov et al. 1997). Comparisons with other dendritic RNAs, and with RNAs localized in other cell types, revealed little significant sequence similarity (Muslimov et al. 1997). On the other hand, computer modeling revealed that the BC1 5' domain forms a stable stem-loop structure (Deininger et al. 1996). An alternative clover-leaf like sec-

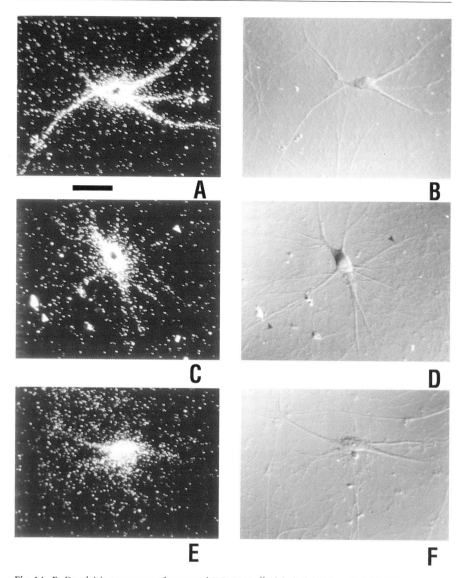

Fig. 1A–F. Dendritic transport of neuronal BC1 RNA. ^{35}S-labeled full-length BC1 RNA was transcribed in vitro and microinjected into somata of sympathetic neurons in primary culture (**A**, **B**). Four hours after injection, significant levels of the injected RNA were observed in dendrites. *Open arrows* in **A** indicate clustered labeling signal over some dendritic segments. **C, D** Nuclear U4 RNA was injected. No significant labeling was observed in medial and distal dendritic segments (*arrowheads*). Proximal segments of some dendrites showed low-level labeling. Similar results were obtained when an RNA of random sequence (144 nt) was injected (**E, F**). **A, C, E** Dark field optics; **B, D, F** Nomarski (DIC) optics. *Scale bar* 100 μm. (Muslimov et al. 1997, with permission)

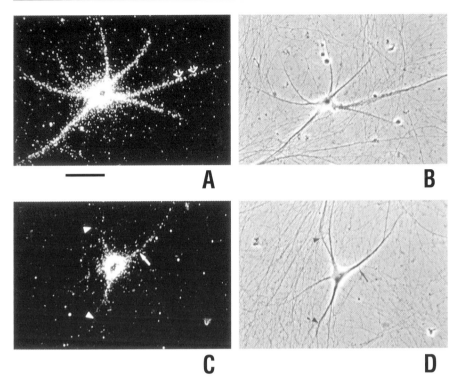

Fig. 2A–D. Differential transport competence of BC1 5′ and 3′ domains. Cell body and dendrites of a neuron injected with the BC1 5′ domain were strongly labeled (**A, B**). *Open arrowheads* indicate clustering of the labeling signal. In contrast, following injection of the BC1 3′ domain (**C, D**), injected cells were labeled over the cell body and proximal-most dendritic segments only (*arrows*). Medial and distal dendritic segments showed no significant labeling (*arrowheads*). **A, C** Dark field optics; **B, D** phase-contrast optics. *Scale bar* 100 µm. (Muslimov et al. 1997, with permission)

ondary structure was of lower predicted stability and was not found to be compatible with recent results from chemical probing experiments (unpublished). We therefore presume that the dendritic targeting competence of BC1 RNA is determined by primary and/or secondary structure elements within a stem-loop 5′ domain. While it remains to be seen to what extent the *cis*-acting BC1 targeting element (BTE) is analogous to targeting elements likely to be contained in other dendritic RNAs, it should be noted that different dendritic RNAs may be targeted to different destination points in dendrites. We suggest that such differential targeting is achieved by use of modular motifs such that the ultimate dendritic destination of a given RNA is specified by the particular choice and combination of such motifs that make up its *cis*-acting targeting element.

While RNA localization to dendrites has by now been amply documented, the issue of RNAs in axons has remained controversial (reviewed by Tiedge et al. 1999). Earlier reports have indicated that axons of lower vertebrates and invertebrates contain ribosomes and RNA, indicating the potential for local axonal protein synthesis (reviewed by Van Minnen 1994). In contrast, the capacity for local translation in mammalian axons may be limited to developing or specialized axonal systems (Tiedge et al. 1993; Ressler et al. 1994; Vassar et al. 1994; Bassell et al. 1998). This raises the question as to whether axonal RNA transport is more widespread in axons of invertebrates and lower vertebrates, compared with mature mammalian axons. BC1 RNA is transported in hypothalamo-neurohypophyseal axons (Tiedge et al. 1993), but little or no BC1 RNA was detectable in other axons in the rat nervous system (Tiedge et al. 1991; Muslimov et al. 1998). We therefore asked whether BC1 RNA is recognized and transported in axons of lower vertebrates. In collaboration with E. Koenig (SUNY Buffalo), we microinjected Mauthner neurons in live goldfish with various RNAs. We found that full-length BC1 RNA and the BC1 5' domain were specifically and rapidly targeted to Mauthner cell dendrites and axons, whereas the BC1 3' domain, nuclear U4 and U6 RNAs, and irrelevant RNAs of similar length were not transported (unpubl.). From time course experiments, we calculated an average transport rate of 2 µm/s, a velocity that would qualify as fast axonal transport. Thus, a targeting element in BC1 RNA that is responsible for dendritic transport in mammalian neurons seems to be recognized by dendritic and axonal transport systems in Mauthner neurons. These results indicate that RNA transport in nerve cells is a mechanism that arose early in evolution and has been conserved since. It is conceivable that such mechanisms were in fact first established in axons, when dendrites were still rather rudimentary: in modern mammalian neurons with complex, elaborate dendrites, it appears that the biological significance of dendritic RNA targeting may have eclipsed that of axonal targeting.

3 Activity-Dependent Regulation

Dendritic RNAs have been implicated in growth and plasticity of synapses (Steward 1997). Long-term modulation of synaptic form and function through dendritic RNAs would necessitate a feedback loop in which at least some of these RNAs are subject to activity-dependent regulation. We have therefore examined the regulation of BC1 RNA as a function of (1) synapse formation and (2) neuronal activity.

Fig. 3A–I. Modulation of somatodendritic BC1 expression by neuronal activity. **A–C** Hippocampal neurons were grown in primary culture for 14 days. **D–F** Hippocampal neurons were grown for 14 days in the presence of 1 µm TTX. **G–I** Hippocampal neurons were grown for 9 days in the presence of 1 µm TTX, and for a further 5 days in the absence of TTX. *Left column* BC1 RNA, dark field optics; *middle column* BC1 RNA, phase-contrast optics, corresponding to **A, D, G**; *right column* 7SL RNA, dark field optics. **D** was overexposed to demonstrate the absence of any significant labeling over cell bodies and dendrites. *Scale bar* 50 µm. (Muslimov et al. 1998, with permission)

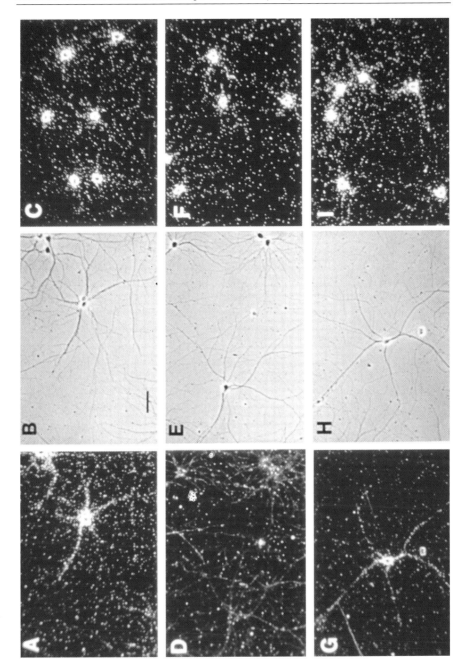

Working with hippocampal neurons developing in primary culture, we established that the onset of BC1 expression is coordinated with the initial formation of synapses (Muslimov et al. 1998). Delayed synapse formation resulted in an analogously delayed onset of BC1 expression, indicating that the former is a prerequisite for the latter. What, then, is the trigger for BC1 expression – the physical formation of synaptic contacts or rather *trans*-synaptic activity conducted through them? To address this question, we used tetrodotoxin (TTX) to inhibit electrical activity in hippocampal neurons in culture. Such inhibition resulted in a significant but fully reversible downregulation of somatodendritic BC1 levels (Fig. 3; see Muslimov et al. 1998). This downregulation was specific, since levels of 7SL RNA, the RNA component of the signal recognition particle (Walter and Blobel 1982), were not affected. These results thus show that it is not the mere physical contact but rather the physiological activity of a synapse that is responsible for inducing expression of BC1 RNA.

It is likely that the observed modulation of somatodendritic BC1 levels by neuronal activity is achieved through regulation of transcription in the nucleus, although additional mechanisms, such as modulation of RNA stability or regulation of delivery to dendrites, cannot be ruled out at this time. The consequences of gene regulation in the nucleus are in principle global: increased levels of BC1 RNA would be available cell-wide. Obviously, the effects of increased cell-wide levels of BC1 RNA may still be synapse-specific as newly synthesized RNA may be preferentially recruited by activated synapses. In this scenario, an activated synapse would be identified by some modification ('tag') that would promote interaction of newly synthesized BC1 RNA with that synapse (Tiedge et al. 1999). Alternatively, higher amounts of BC1 RNA may be delivered cell-wide to all synapses, but the function of BC1 RNA at any given synapse may be dependent on previous and/or future activity at that synapse.

4 Outlook

We have previously suggested that BC1 RNA functions at the interface between mRNA and protein synthetic machinery in dendrites (Muslimov et al. 1998). BC1 RNA is one of the most prominent RNA components in postsynaptic microdomains such as dendritic spines (Chicurel et al. 1993; Rao and Steward 1993), and we interpret its colocalization in such microdomains with dendritic mRNAs and components of the translational apparatus to indicate an adapter role in dendritic translation. In such a role, BC1 RNA would interact with dendritic mRNAs (or a subset of such mRNAs) and the local protein synthetic machinery in the activity-dependent modulation of dendritic translation. It is also possible that BC1 RNA plays a role in dendritic RNA transport, although such a role would seem less likely since dendritic localization of selected mRNAs appears to be normal in BC1-negative mice (Skryabin et al., unpubl. results). However, it cannot be ruled out at this point that developmental compensations may have obscured any phenotype changes related to dendritic RNA transport.

In summary, our current view of BC1 RNA is that of a translational modulator in postsynaptic microdomains. Previous work has demonstrated active protein synthesis, and its activity-dependent modulation, in dendrites (Torre and Steward 1992; Wu et al. 1998; Ouyang et al. 1999; Steward and Halpain 1999). Our conjecture of BC1 RNA as a modulator of local protein synthesis is a testable hypothesis that we hope to examine in future work.

Acknowledgements. Work in the authors' laboratories was supported by the Deutsche Forschungsgemeinschaft (J.B.) and NIH (H.T.).

References

Ainger K, Avossa D, Morgan F, Hill SJ, Barry C, Barbarese E, Carson JC (1993) Transport and localization of exogenous myelin basic protein mRNA microinjected into oligodendrocytes. J Cell Biol 123:431–441

Ainger K, Avossa D, Diana AS, Barry C, Barbarese E, Carson JC (1997) Transport and localization elements in myelin basic protein mRNA. J Cell Biol 138:1077–1087

Bassell GJ, Zhang H, Byrd AL, Femino AM, Singer RH, Taneja KL, Lifshitz LM, Herman IM, Kosik KS (1998) Sorting of β-actin mRNA and protein to neurites and growth cones in culture. J Neurosci 18:251–265

Bassell GJ, Oleynikov Y, Singer RH (1999) The travels of mRNAs through all cells large and small. FASEB J 13:447–454

Brosius J (1999a) RNAs from all categories generate retrosequences that may be exapted as novel genes or regulatory elements. Gene 238:115–134

Brosius J (1999b) Genomes were forged by massive bombardments with retroelements and retrosequences. Genetica 107:209–238

Brosius J, Gould SJ (1992) On 'genomenclature', a comprehensive (and respectful) taxonomy for pseudogenes and other 'junk DNA'. Proc Natl Acad Sci USA 89:10706–10710

Brosius J, Gould SJ (1993) Molecular constructivity. Nature 365:102

Brosius J, Tiedge H (1995) Neural BC1 RNA: dendritic localization and transport. In: Lipshitz HD (ed) Localized RNAs. Landes, Austin, pp 289–300

Cheng J-G, Tiedge H, Brosius J (1996) Identification and characterization of BC1 RNP particles. DNA Cell Biol 15:549–559

Chicurel ME, Terrian DM, Potter H (1993) mRNA at the synapse: analysis of a preparation enriched in hippocampal dendritic spines. J Neurosci 13:4054–4063

DeChiara TM, Brosius J (1987) Neural BC1 RNA: cDNA clones reveal nonrepetitive sequence content. Proc Natl Acad Sci USA 84:2624–2628

Deininger PL, Tiedge H, Kim J, Brosius J (1996) Evolution, expression, and possible function of a master gene for amplification of an interspersed repeated DNA family in rodents. In: Cohn W, Moldave K (eds) Progress in nucleic acid research and molecular biology, vol 52. Academic Press, San Diego, pp 67–88

Gardiol A, Racca C, Triller A (1999) Dendritic and postsynaptic protein synthetic machinery. J Neurosci 19:168–179

Gould SJ, Vrba ES (1982) Exaptation a missing term in the science of form. Paleobiology 8:4–15

Kobayashi S, Goto S, Anzai K (1991) Brain-specific small RNA transcript of the identifier sequences is present as a 10S ribonucleoprotein particle. J Biol Chem 266:4726–4730

Kuhl D, Skehel P (1998) Dendritic localization of mRNAs. Curr Opin Neurobiol 8:600–606

Lu C, Bablanian R (1996) Characterization of small nontranslated polyadenylated RNAs in vaccinia virus-infected cells. Proc Natl Acad Sci USA 93:2037–2042

Martignetti JA, Brosius J (1993) Neural BC1 RNA as an evolutionary marker: guinea pig remains a rodent. Proc Natl Acad Sci USA 90:9698–9702

Martignetti JA, Brosius J (1995) BC1 RNA: transcriptional analysis of a neural cell-specific RNA polymerase III transcript. Mol Cell Biol 15:1642–1650

Mayford M, Baranes D, Podsypanina K, Kandel ER (1996) The 3′-untranslated region of CaMKIIα is a *cis*-acting signal for the localization and translation of mRNA in dendrites. Proc Natl Acad Sci USA 93:13250–13255

Muslimov IA, Santi E, Homel P, Perini S, Higgins D, Tiedge H (1997) RNA transport in dendrites: a *cis*-acting targeting element is contained within neuronal BC1 RNA. J Neurosci 17:4722–4733

Muslimov IA, Banker G, Brosius J, Tiedge H (1998) Activity-dependent regulation of dendritic BC1 RNA in hippocampal neurons in culture. J Cell Biol 141:1601–1611

Ouyang Y, Rosenstein A, Kreiman G, Schuman EM, Kennedy MB (1999) Tetanic stimulation leads to increased accumulation of Ca^{2+}/calmodulin-dependent protein kinase II via dendritic protein synthesis in hippocampal neurons. J Neurosci 19:7823–7833

Rao A, Steward O (1993) Evaluation of RNAs present in synaptodendrosomes: dendritic, glial, and neuronal cell body contribution. J Neurochem 61:835–844

Ressler KJ, Sullivan SL, Buck LB (1994) Information coding in the olfactory system: evidence for a stereotyped and highly organized epitope map in the olfactory bulb. Cell 79:1245–1255

Steward O (1997) mRNA localization in neurons: a multipurpose mechanism? Neuron 18:9–12

Steward O, Halpain S (1999) Lamina-specific synaptic activation causes domain-specific alterations in dendritic immunostaining for MAP2 and CAM kinase II. J Neurosci 19:7834–7845

Taylor BA, Navin A, Skryabin BV, Brosius J (1997) Localization of the mouse gene (*Bc1*) encoding neural BC1 RNA near the fibroblast growth factor 3 locus (*Fgf3*) on distal chromosome 7. Genomics 144:153–154

Tiedge H, Brosius J (1996) Translational machinery in dendrites of hippocampal neurons in culture. J Neurosci 16:7171–7181

Tiedge H, Fremeau RT Jr, Weinstock PH, Arancio O, Brosius J (1991) Dendritic location of neural BC1 RNA. Proc Natl Acad Sci USA 88:2093–2097

Tiedge H, Dräger UC, Brosius J (1992) Murine BC1 RNA in dendritic fields of the retinal inner plexiform layer. Neurosci Lett 141:136–138

Tiedge H, Zhou A, Thorn NA, Brosius J (1993) Transport of BC1 RNA in hypothalamo-neurohypophyseal axons. J Neurosci 13:4114–4219.

Tiedge H, Bloom FE, Richter D (1999) RNA, wither goest thou? Science 283:186–187

Torre ER, Steward O (1992) Demonstration of local protein synthesis within dendrites using a new cell culture system which permits the isolation of living axons and dendrites from their cell bodies. J Neurosci 12:762–772

Torre ER, Steward O (1996) Protein synthesis within dendrites: glycosylation of newly synthesized proteins in dendrites of hippocampal neurons in culture. J Neurosci 16:5967–5978

Van Minnen J (1994) RNA in the axonal domain: a new dimension in neuronal functioning? Histochem J 26:377–391

Vassar R, Chao SK, R. S, Nunez JM, Vosshall LB, Axel R (1994) Topographic organization of sensory projections to the olfactory bulb. Cell 79:981–991

Vidal F, Cuzin F (1995) SINE-derived motifs and the regulation of RNA polymerase II transcripts. In: Maraia R (ed) The impact of short interspersed elements (SINEs) on the host genome. Landes, Austin, pp 125–131

Walter P, Blobel G (1982) Signal recognition particle contains a 7 S RNA essential for protein translocation across the endoplasmic reticulum. Nature 299:691–698

Weiner AM, Deininger PL, Efstratiadis A (1986) Nonviral retroposons: genes, pseudogenes, and transposable elements generated by the reverse flow of genetic information. Annu Rev Biochem 55:631–661

Wu L, Wells D, Tay J, Mendis D, Abbott MA, Barnitt A, Quinlan E, Heynen A, Fallon JR, Richter JD (1998) CPEB-mediated cytoplasmic polyadenylation and the regulation of experience-dependent translation of α-CaMKII mRNA at synapses. Neuron 21:1129–1139

Nucleocytoplasmic mRNA Transport

Yingqun Huang[2] and Gordon G. Carmichael[1]

1 Introduction

How do messenger RNA molecules (mRNAs) migrate from the nucleus to the cytoplasm? Messenger RNA precursors (pre-mRNAs) are synthesized in the nucleus by RNA polymerase II and are then subjected to a series of processing reactions which include the addition of a 7-methylguanosine cap at the 5′ end, the removal of introns by splicing, and the generation of mature 3′ ends, usually by polyadenylation. Mature mRNA molecules are then translocated to the cytoplasm through the nuclear pore complex. This process must be selective, because unprocessed pre-mRNAs are not exported. Although the detailed mechanism of mRNA nuclear export is not yet well understood, recent work has begun to lay a useful framework for understanding the nature and regulation of this fundamental biological process. This chapter will describe the essential features of mRNA export. Due to space limitations, much important and interesting work on mRNA transport in viral model systems will not be included. However, a number of good recent reviews on mRNA export have been published (Corbett and Silver 1997; Nakielny et al. 1997; Nigg 1997; Izaurralde and Adam 1998; Ohno et al. 1998; Stutz and Rosbash 1998; Adam 1999).

2 The Export Machinery

2.1 Nuclear Pores

Most macromolecules do not pass freely between the nucleus and cytoplasm. This is intimately related to the structure of the nuclear pore complex (NPC). Vertebrate NPCs are extremely large, roughly cylindrical structures, with a mass of about 125 MDa (Rout and Wente 1994; Doye and Hurt 1997). They consist of two membrane-spanning rings and contain aqueous channels through which molecules smaller than 40–60 kDa can freely pass. There is a

[1] Department of Microbiology, University of Connecticut Health Center, Farmington, CT 06030, USA
[2] Y. Huang, Department of Molecular Biophysics and Biochemistry, Howard Hughes Medical Institute, Yale University School of Medicine, 295 Congress Avenue, New Haven, CT 06536, USA

Results and Problems in Cell Differentiation, Vol. 34
D. Richter (Ed.): Cell Polarity and Subcellular RNA Localization
© Springer-Verlag Berlin Heidelberg 2001

central plug, or transporter, through which export of larger substrates occurs. Conformational changes in the central plug allow large particles to pass through it. The major protein components of the NPC are known as nucleo-porins. The NPC is composed of 50–100 different nucleoporins, only a fraction of which have been identified. There are two main classes: integral membrane proteins (the NPC core) and FG (phenylalanine and glycine) repeat-rich pro-teins which are for the most part peripheral (Davis 1995; Doye and Hurt 1997; Stutz and Rosbash 1998). On the cytoplasmic side of the pore there are eight filaments that protrude 30–50 nm away from the nuclear envelope. Inside the nucleus is an attached structure called the nuclear basket which extends about 100 nm into the nucleoplasm. The cytoplasmic filaments and nuclear basket almost certainly play important roles in nucleocytoplasmic protein and RNA trafficking, but these roles have not yet been determined. Since the NPC is so large, transport of macromolecules between the nuclear and cytoplasmic com-partments involves numerous interactions of transport cargo and associated factors with components of the NPC. Transport through the NPC may consist of a series of association-dissociation reactions that together act to propel a transport substrate throughout the length of the NPC. Energy must be involved in this process, but how energy is used remains unclear at the molecular level.

2.2 Import and Export Receptors

Facilitated nuclear import or export must involve the interaction of cargo molecules with receptors or adapters. Cargo is moved from one subcellular compartment to another via factors that contain specific trafficking signals. These signals are recognized by transport receptors in one compartment. After transport, the receptors are released and recycled to the original compartment. Both import and export receptors shuttle between the nucleus and the cyto-plasm (Görlich and Mattaj 1996; Görlich et al. 1997; Nigg 1997). Thus, for mRNA export, specific receptors must accompany the messages from the nucleus to the cytoplasm. To be reused, these proteins have to return to the nucleus.

For import into the nucleus, there are two classical types of nuclear local-ization signal (NLS): a simple stretch of three to five basic amino acids, often with a proline or glycine; and a bipartite signal of a ten amino acid sequence downstream of a basic dipeptide (Dingwall and Laskey 1991). The NLS-containing protein is recognized by an import receptor, which is a heterodimer of two polypeptides, importin-α and importin-β (Görlich et al. 1996; Nigg 1997). This trimeric complex then appears to dock at the NPC, perhaps to the nucleoporin RanBP2 (Melchior et al. 1995; Wu et al. 1995; Yokoyama et al. 1995). An important feature of this complex is that importin-α binds to the NLS-containing protein (which is the import cargo), and importin-β binds to importin-α as well as to the NPC. Thus, import would involve multiple, simul-taneous protein-protein contacts. Different classes of import cargoes might interact with a family of importin-α-like proteins, which all interact with importin-β for import (Görlich 1997). In addition to the receptors for proteins with classical NLS elements, there have also been reports of a family of importin-β-like proteins that mediate the import of specific classes of nuclear proteins (Pollard et al. 1996; Fridell et al. 1997; Görlich et al. 1997).

Nuclear export signals (NES elements) have also been described, and might function in a conceptually analogous way to NLS elements. Many or most proteins that localize to the nucleus may contain NLS elements. Proteins that mediate mRNA export must also contain NES elements to allow these proteins to both enter and leave the nucleus. Specific sequences that act as NES elements have now been described for a number of nuclear proteins, including hnRNP A1 (Michael et al. 1995a; Siomi and Dreyfuss 1995), hnRNP K (Michael et al. 1995b) and the HIV-1 viral protein Rev (Cullen and Malim 1991). The Rev protein binds to a specific sequence in viral RNAs, the Rev-responsive element, or RRE (Malim and Cullen 1991). Binding may directly lead to the export of RNAs containing the RRE (Fischer et al. 1994). The Rev NES is a short leucine-rich stretch of amino acids: $L-X_{2-3}-Y-X_{2-3}-L-X-L/I$, where X is any amino acid and Y is L/I/F/V or M (Bogerd et al. 1996). A domain like this exists in several known proteins, including TFIIIA (Fischer et al. 1995; Fridell et al. 1996) and two yeast pore-associated proteins, Gle1p (Murphy and Wente 1996), and Mex67p (Segref et al. 1997).

NES-containing proteins are recognized in the nucleus by export receptors, called exportins. The first such receptor described was a protein called CRM1/exportin 1, which acts as the export receptor for the Rev NES (Fornerod et al. 1997a; Stade et al. 1997). Although the Rev protein can mediate the export of unspliced or partially spliced HIV-1 viral RNAs (see below), CRM1 does not appear to act as the export receptor for most cellular mRNAs, but rather is important for U snRNA export (Stade et al. 1997).

2.3 The Role of Ran in the Establishment of Compartment Identity

For both import and export, however, the association and dissociation of import/export receptors must be mediated by additional components that serve to identify which compartment the complexes are in. Directional transport requires that the import or export machinery can distinguish clearly between the nucleus and the cytoplasm at all times. Compartmental identity is established by the small GTPase Ran (Fig. 1). Ran is an extremely abundant and predominantly nuclear protein that exists in two forms. In the nucleus, almost all of the Ran is bound to GTP (RanGTP). In the cytoplasm, almost all of the Ran is bound to GDP (RanGDP). Thus, there is a RanGTP/RanGDP gradient across the NPC. As for other small GTPases within the cell, the nucleotide-bound state of Ran is regulated by additional cofactors. In the cytoplasm, the intrinsic GTPase activity of Ran is used to convert RanGTP to RanGDP. This intrinsic GTPase activity is stimulated by up to five orders of magnitude by the protein RanGAP1 (Ran GTPase-activating protein-1; Bischoff et al. 1994, 1995), and additionally tenfold by RanBP1 (Ran-binding

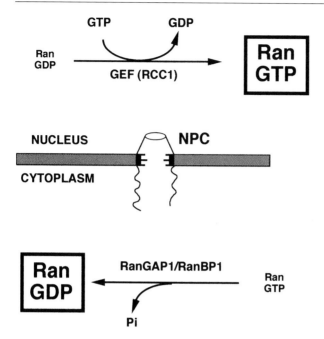

Fig. 1. Ran is used to establish compartment identity. RanGTP levels are high in the nucleus, while RanGDP levels are higher in the cytoplasm. In the nucleus the chromosome-associated guanine nucleotide exchange factor GEF(RCC1) maintains Ran in the GTP-bound state. In the cytoplasm, GTP hydrolysis is favored, and is stimulated by RanGAP1 and RanBP1. See text for details

protein 1; Coutavas et al. 1993). Both RanGAP1 and RanBP1 are found only in the cytoplasm (Hopper et al. 1990; Melchior et al. 1993; Bischoff et al. 1995; Richards et al. 1996). In the nucleus the situation is quite different. In this compartment, Ran is maintained predominantly in the GTP-bound state by the action of a chromatin-bound guanine nucleotide exchange factor, GEF (also called RCC1). This factor catalyzes the exchange of GTP for GDP.

How does the nucleotide bound state of Ran control import and export? One important way that the state of Ran influences nucleocytoplasmic transport is through interactions with the adapter proteins that promote import or export. For example, import complexes form in the presence of RanGDP in the cytoplasm (Görlich 1997 no. 5728), while export complexes in the nucleus form only in the presence of RanGTP, and disassemble in the presence of RanGDP (Fornerod et al. 1997a; Kutay et al. 1997; Stade et al. 1997). Thus, for a process such as mRNA export, functional complexes are assembled in the nucleus, translocated to the cytoplasm, and then disassembled, allowing the recycling of the receptors and adapters. Surprisingly, the translocation step through the NPC does not appear to require the hydrolysis of RanGTP, so the source of energy for transport remains obscure (Nakielny and Dreyfuss 1998; Schwoebel et al. 1998; Englmeier et al. 1999).

Figure 2 shows a simple model for the role of export receptors, adapters and RanGTP in RNA export. An example of this model for mRNA export is that mediated by the HIV-1 Rev protein, whose NES receptor is CRM1. CRM1 is a Ran binding protein that shares similarities with the Ran-binding domain of importin-β (Fornerod et al. 1997a; Görlich et al. 1997). In vitro, CRM1 forms a complex with RanGTP and various NES sequences and the association is cooperative and inhibited by the drug leptomycin B (Fornerod et al. 1997a). The model for CRM1 function is that the NES-protein binds RNA in the nucleus, and this complex then associates with CRM1, but only in the presence of RanGTP (Stade et al. 1997). Multiple interactions of this type may occur along the length of the RNA molecule. The large complex is then translocated through the pore via sequential interactions with nucleoporins (Fritz and Green 1996; Stutz et al. 1996). In the cytoplasm, hydrolysis of GTP leads to NES/cargo release, because CRM1/NES interaction requires RanGTP. CRM1 would then recycle to the nucleus to be used again. Floer and Blobel (1999) have recently uncovered several new interactions that shed light on CRM1 function in nuclear protein (and perhaps also mRNA) export. First, the Rev/CRM1/RanGTP complex interacts with some nucleoporins, but not others. These interactions lead to the release of Rev, leaving aNup/CRM1/RanGTP complex. Second, RanBP1 can displace Nup and form a ternary RanBP1/RanGTP/CRM1 complex, that can be dissolved by GTP hydrolysis mediated by RanGAP. Finally, the RanBP1/RanGTP/CRM1 complex is dissociated to recycle the Ran/RanGEF complex and allow reformation of a Rev/CRM1/RanGTP complex.

3 Cellular mRNA Export

Messenger RNA molecules that are exported to the cytoplasm are typically capped at their 5'-ends, polyadenylated at their 3'-ends, and are generally processed from larger precursors by pre-mRNA splicing. RNAs in the nucleus exist as ribonucleoprotein (RNP) complexes. Also, not all RNAs are chosen for export. Therefore, signals in the RNAs must contribute to export, along with factors that bind to these molecules. *cis*-Acting signals include the monomethyl cap structure at the 5'-end, the poly(A) tail at the 3'-end and sequences within the body of the message. Proteins that interact with these elements might be key players in the mRNA export process, and might mediate interactions with mRNA-specific exportins. Does mRNA transport use an exportin? While highly likely, so far, there is no direct evidence that this is so. In higher eukaryotes, CRM1 probably plays little role in mRNA export.

3.1 The Cap Structure

mRNAs contain a monomethylated m^7G cap structure at their 5'-ends, which is added shortly after transcription initiation. There is a nuclear cap-binding complex (CBC) that consists of two proteins, CBP80 and CBP20 (Ohno et al.

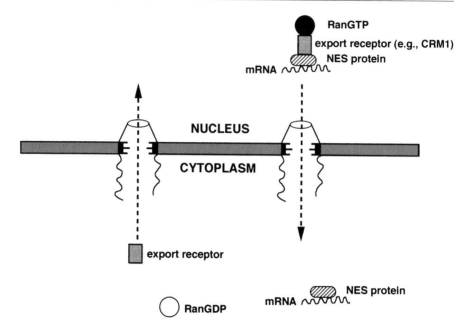

Fig. 2. A model for the role of export receptors, adapters, and RanGTP in mRNA export. mRNA is the cargo, and is recognized by an RNA-binding protein containing an RNA export signal (NES). This protein, in turn, is recognized by an export receptor (for example, CRM1 when the NES protein is HIV-1 Rev). This interaction occurs only in the presence of RanGTP, and the complex of all four components is the export substrate, which interacts with the NPC and is translocated to the cytoplasm. In the cytoplasm, RanGTP is hydrolyzed to RanGDP, leading to dissociation of the export receptor from the mRNA cargo, and recycling of the export receptor to the nucleus

1990; Izaurralde et al. 1994; Kataoka et al. 1994). Microinjection experiments using *Xenopus* oocytes suggested that the cap structure and its associated proteins play a positive, but not absolutely essential role, in mRNA export (Hamm and Mattaj 1990; Izaurralde et al. 1992, 1995; Jarmolowski et al. 1994).

3.2 The Poly(A) Tail

Unlike the cap at the 5′-end of the message, proper 3′-end formation, usually the addition of a poly(A) tail, may be required for mRNA export (Eckner et al. 1991; Huang and Carmichael 1996a). Although *Xenopus* oocyte microinjection studies suggested that polyadenylation is not essential for export, mRNAs produced from plasmids introduced into cells by transfection cannot be exported to the cytoplasm unless their 3′-ends are properly processed by the cellular

machinery (Huang and Carmichael 1996a). This requirement cannot be overcome using the HIV-1 Rev protein and its RRE element, or by encoding a
poly(A) tail in the transcribed sequence (Huang and Carmichael 1996a). The
process of polyadenylation in the nucleus requires a number of cis-acting RNA
elements and trans-acting factors (reviewed by Colgan and Manley 1997; Wahle
and Kühn 1997; Zhao et al. 1999). Several of these factors may remain bound
to the mRNA molecule after processing. The cleavage and polyadenylation
specificity factor, CPSF, binds upstream of the poly(A) tail and has not been
shown to be released after polyadenylation. Likewise, poly(A)-binding protein
II (PABP II) is intimately involved in regulating poly(A) tail length and is tightly
associated with nuclear polyadenylated messages (Wahle 1991; Bienroth et al.
1993). The possibility remains that either of these factors might play an important role in mRNA export. Perhaps importantly, PABP II is not found in the
cytoplasm. Rather, the poly(A) tail of cytoplasmic mRNA is covered by poly(A)
binding protein I (PABP I; Gorlach et al. 1994). It remains to be determined how
and when the poly(A) tail is remodeled, and whether the replacement of PABP
II with PABP I is intimately associated with mRNA export.

3.3 hnRNP Proteins

Likely candidates for essential mRNA export factors would be abundant
nuclear RNA-binding proteins. The most abundant nuclear proteins of this
type are the hnRNP proteins. hnRNP proteins associate with mRNA during
or shortly after transcription, and most of these shuttle between the nucleus
and the cytoplasm (Nakielny and Dreyfuss 1997). There are at least 20
characterized hnRNP proteins in vertebrate cells (Dreyfuss et al. 1993). The
only ones shown not to shuttle so far are hnRNP C and hnRNP U (Piñol-
Roma and Dreyfuss 1992, 1993; Michael et al. 1995a; Nakielny and Dreyfuss
1996).

   The best-studied hnRNP protein is hnRNP A1, and it serves as a useful
model for the potential role of hnRNP proteins in mRNA export. This protein
has a domain (the M9 domain) which serves both as an NLS and an NES
(Michael et al. 1995b). A1 appears to remain bound to mRNAs during translocation from the nucleus to the cytoplasm. Injection of a large excess of A1 into
Xenopus oocytes inhibited mRNA export (Izaurralde et al. 1997), but injection
of the M9 domain alone did not inhibit export (Pollard et al. 1996; Izaurralde
et al. 1997). This suggested that A1 contains an element or domain that can
compete for a receptor involved in the export of many cellular messages.
However, export of some mRNAs was not inhibited by excess A1, suggesting
multiple mechanisms or receptors (Saavedra et al. 1997). Consistent with this
is the observation that hnRNP K has a different shuttling signal (Michael et al.
1997). Export must involve not only RNA-binding proteins that contain NES
elements, but at least one, and possibly many, export receptors, yet to be
discovered.

It is likely that different primary transcripts are decorated by different sets or patterns of hnRNP proteins, each of which displays a unique set of sequence preferences for RNA binding (Matunis et al. 1993). RNAs that are bound by nonshuttling hnRNPs such as hnRNP C or hnRNP U might be retarded or blocked in their export. Alternatively, these nonshuttling proteins may be removed from mRNPs before export occurs. Finally, hnRNP binding alone might not be sufficient for export. Recent work in the yeast system has suggested that arginine methylation of hnRNP proteins may be required for mRNA export (Shen et al. 1998). Cells genetically deficient in an hnRNP methylase produced hnRNP proteins that could no longer shuttle between the nucleus and the cytoplasm.

3.4 SR Proteins

SR proteins belong to a conserved class of related and abundant nuclear proteins that are both essential for constitutive pre-mRNA splicing and are involved in the regulation of alternative splicing (Ge and Manley 1990; Krainer et al. 1990; Zahler et al. 1993; Fu 1995; Manley and Tacke 1996; Cáceres et al. 1997). This activity can be antagonized by some members of the hnRNP protein family in a relative concentration-dependent manner (Mayeda and Krainer 1992; Mayeda et al. 1994). SR proteins are nuclear phosphoproteins that contain repeated stretches of serine and arginine residues (RS domains). Their RNA binding sites have most often been found within exons (Fu 1995; Cáceres et al. 1997). In mammalian cells, a subset of SR proteins appear to remain bound to the mature mRNA after in vitro splicing (Blencowe et al. 1994, 1995). Therefore, the possibility exists that at least some of them might accompany mRNAs from the nucleus to the cytoplasm. Recent work has shown that some, but not all, SR proteins are shuttling proteins, indicating that different members of this family might have different functions (Cáceres et al. 1998). For example, SRp20, SF2/ASF and 9G8 shuttle, while SC35 and SRp40 do not. The RS domains are necessary but not sufficient for shuttling. RNA binding appears to be necessary, but it is not known whether RNA binding confers the shuttling property. Also, increased levels of SR protein phosphorylation seemed to lead to greater levels in the cytoplasm, suggesting a role for phosphorylation in shuttling. Recent results suggest that association of SR proteins with specific SR protein kinases might mediate transport (Koizumi et al. 1999).

3.5 Visualization of mRNA Export

Some important lessons in mRNA export have come from elegant electron microscopic studies reported by Daneholt and colleagues (Daneholt 1997). This group has investigated the synthesis, processing and export of giant Balbiani Ring transcripts from the salivary glands of *Chironomous tentans*. The RNP particles appear to be assembled into ring-like structures, with their

5′ and 3′ ends relatively close together (Mehlin et al. 1995). Both hnRNP and SR proteins are associated with them. Incorporating these observations with those described above leads to a generalized model for mRNP structure presented in Fig. 3A.

Visa et al. (1996b) showed that CBC accompanies Balbiani ring (BR) mRNA out of the nucleus. Also, an hnRNP protein homologous to hnRNP A1 not only bound along the length of transcripts within the nucleus, but accompanied the BR mRNA through the NPC and into cytoplasmic polysomes (Visa et al. 1996a). In another study, an insect SR protein was shown to bind to BR RNP particles within the nucleus, but not to accompany the mRNA to the cytoplasm (Alzhanova-Ericsson et al. 1996). Finally, recent work has allowed visualization of the export process of these mRNPs (Daneholt 1997). The mRNP ring-like structure appears to associate first with the nuclear basket. Then, the 5′-end enters the NPC. As the mRNP is translocated through the nuclear pore in a 5′ to 3′ direction, the RNP unwinds or is remodeled. In fact, there have been reports of enzymatic activities at or near the pore that could contribute to RNP remodeling as well as RNP export (Snay-Hodge et al. 1998; Tseng et al. 1998; Hodge et al. 1999; Schmitt et al. 1999). One of these activities is an RNA helicase which might participate directly in RNP remodeling or translocation (Schmitt et al. 1999). On the cytoplasmic side of the pore, ribosomes may attach to the mRNA even before the entire message has traversed the pore (Daneholt 1997). This and several other observations described above lead to the generalized model for export presented in Fig. 3.

3.6 The Role of Splicing in mRNA Export

For intron-containing genes, mRNA export usually does not occur unless introns are removed by splicing. Many intronless transcripts encoded by cDNAs of intron-containing genes are not exported (Gruss et al. 1979; Hamer and Leder 1979; Buchman and Berg 1988; Ryu and Mertz 1989). Two models have been suggested. The first hypothesizes that splicing and export may be coupled processes and that splicing may be a prerequisite for export. One way this coupling could occur would be that positive export factors are deposited via splicing. The second, alternative model, suggests that unspliced transcripts may be retained in the nucleus because of association of spliceosomes (Legrain and Rosbash 1989). Studies from the polyoma virus model system suggested that spliceosome retention may indeed play an important role – release from splicing factors that interact with 5′ splice sites allowed export of mRNAs, whether or not they were spliced (Huang and Carmichael 1996b).

Two other major issues have arisen with respect to splicing and mRNA export. The first is that some intron-containing viral mRNAs can be exported without splicing. The best-studied mRNAs of this sort are retroviral messages. Retroviruses encode elements that can override nuclear retention and allow

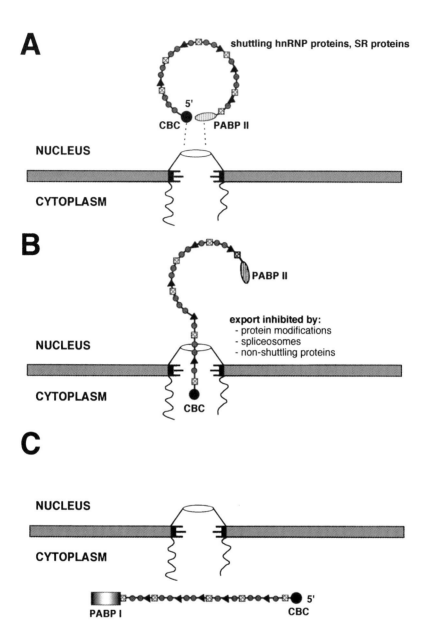

Fig. 3A–C. A model for mRNA export. **A** Within the nucleus mRNPs consist of the mRNA, CBC, PABP II, hnRNP proteins, SR proteins, and perhaps additional factors. The 5'-end is positioned close to the 3'-end (according to electron microscopic data from the laboratory of B. Daneholt (see text for details). This particle docks first at or near the nuclear basket. Since both CBC and the poly(A) tail are important for export, perhaps both ends of the message need to be recognized simultaneously in order for efficient export to occur. This would ensure that only complete, intact mRNAs are translocated. **B** Translocation through the NPC occurs, beginning with the 5'-end. During translocation, the mRNP particle is unfolded or remodeled, and translocation involves a series of protein-protein association-dissociation reactions that propel the mRNA into the cytoplasm. Translocation cannot occur if spliceosomes or non-shuttling proteins remain bound to the mRNA, and might be influenced by post-translational modifications to the bound proteins. **C** In the cytoplasm, the mRNA remains associated with a subset of hnRNP proteins, and perhaps SR proteins. PABP I has replaced PABP II on the poly(A) tail

the export of unspliced or partially spliced mRNAs. HIV-1 contains the RRE element which is bound by Rev. After binding to its core sequence, Rev oligomerizes along the RNA (Heaphy et al. 1991; Zapp et al. 1991), and may inhibit spliceosome formation (Chang and Sharp 1989; Dyhrmikkelsen and Kjems 1995). Since Rev contains an NES element, one can envision this protein as a replacement hnRNP protein that is specifically targeted to viral messages in the nucleus. Oligomerization provides the cargo with numerous export signals, and also results in the removal or inhibition of retentive factors (spliceosomes).

Simple retroviruses do not encode a transport protein, but rather appear to have a *cis*-acting element that interacts with cellular factors. Simian type D retrovirus has the constitutive transport element (CTE; Tabernero et al. 1996; Ernst et al. 1997). Substitution of the HIV-1 RRE with the CTE renders HIV-1 replication Rev-independent (Zolotukhin et al. 1994). Like the Rev/RRE interaction, the CTE binds a specific protein. This protein has been identified as TAP (Grüter et al. 1998). TAP has a yeast homologue, Mex67p, which is an essential protein that binds to RNA and to nuclear pores (Santos-Rosa et al. 1998). The human TAP binds to poly(A)$^+$ mRNA (although with rather low affinity), and it is likely that in human cells TAP plays a direct role in mRNA export (Segref et al. 1997). Competition microinjections suggested that TAP is essential for all mRNA export (Pasquinelli et al. 1997; Saavedra et al. 1997). But the pathway is different from Rev (Pasquinelli et al. 1997; Saavedra et al. 1997; Zolotukhin and Felber 1997). Thus, specific interactions between positive *cis*-acting RNA elements and appropriate viral or cellular factors appear to facilitate the cytoplasmic accumulation of intron-containing retroviral mRNAs. However, different retroviruses have chosen different ways to achieve the same effect.

The second issue is that naturally intronless genes can efficiently express cytoplasmic mRNAs (Kedes 1979; Nagata et al. 1980; Hentschel and Birnstiel 1981; Koilka et al. 1987; Hattori et al. 1988), while intronless transcripts from cDNA expression vectors are often not efficiently expressed (Hamer and Leder 1979; Kedes 1979; Nagata et al. 1980; Hentschel and Birnstiel 1981; Hattori et al. 1988; Neuberger and Williams 1988; Ryu and Mertz 1989; Koilka et al. 1987; Jonsson et al. 1992; Nesic et al. 1993). Recently, a growing body of data has suggested that efficient nuclear export of intronless messages is facilitated by specific sequences present within these messages. Examples include the herpes simplex virus thymidine kinase (HSV-TK) message (Liu and Mertz 1995), the hepatitis B virus (HBV) message (Huang and Liang 1993; Huang and Yen 1995) and the mouse histone H2a message (Huang and Carmichael 1997). HnRNP L, which interacts specifically with the HSV-TK element, has been implicated in the nuclear export of HSV-TK messages (Liu and Mertz 1995).

But how do these intronless mRNA transport elements really work? In one recent study, it was found that, in addition to promoting mRNA transport, several of these elements exhibited strong activity not only in inhibition of splicing, but also in stimulation of polyadenylation. Surprisingly, the reported mRNA transport elements from HSV-TK (Liu and Mertz 1995) and HBV (Huang and Liang 1993; Huang and Yen 1995) act in the same manner as the element from histone H2a (Huang and Carmichael 1997; Huang et al. 1999). Moreover, each of these three elements can promote unspliced mRNA export as well as lower the levels of spliced products in an HIV-1 based system (Huang et al. 1999). It was therefore hypothesized that a general feature of transport elements from intronless messages might be to perform three functions: inhibit splicing, enhance mRNA 3'-end processing and promote nuclear export. As we have seen from the results summarized above, each of these processes can affect the cytoplasmic accumulation of mRNA.

4 Conclusions

The export of mRNAs from the nucleus to the cytoplasm is a complex process, with much potential for regulation. The maturation of mRNAs is a dynamic process, in which numerous protein-RNA contacts are made, broken or rearranged. In the process of maturation, mRNAs become associated with proteins that might function in mRNA export. In addition, factors that inhibit export must be removed. Although many of the key factors remain to be identified, a useful conceptual framework has emerged from studies in the past several years.

Acknowledgements. This work was supported by a grant from the National Science Foundation to GGC.

References

Adam SA (1999) Transport pathways of macromolecules between the nucleus and the cytoplasm. Curr Opin Cell Biol 11:402–406
Alzhanova-Ericsson AT, Sun X, Visa N, Kiseleva E, Wurtz T, Daneholt B (1996) A protein of the SR family of splicing factors binds extensively to exonic Balbiani ring pre-mRNA and accompanies the RNA from the gene to the nuclear pore. Genes Dev 10:2881–2893
Bienroth S, Keller W, Wahle E (1993) Assembly of a processive messenger RNA polyadenylation complex. EMBO J 12:585–594
Bischoff FR, Klebe C, Kretschmer J, Wittinghofer A, Ponstingl H (1994) RanGAP1 induces GTPase activity of nuclear ras-related Ran. Proc Natl Acad Sci USA 91:2587–2591
Bischoff FR, Krebber H, Kempf T, Hermes I, Ponstingl H (1995) Human RanGTPase-activating protein RanGAP1 is a homologue of yeast Rna1p involved in mRNA processing and transport. Proc Natl Acad Sci USA 92:1749–1753
Blencowe BJ, Nickerson JA, Issner R, Penman S, Sharp PA (1994) Association of nuclear antigens with exon-containing splicing complexes. J Cell Biol 127:593–607
Blencowe BJ, Issner R, Kim J, McCaw P, Sharp PA (1995) New proteins related to the Ser-Arg family of splicing factors. RNA 1:852–865
Bogerd HP, Fridell RA, Benson RE, Hua J, Cullen BR (1996) Protein sequence requirements for function of the human T-cell leukemia virus type I Rex nuclear export signal delineated by a novel in vivo randomization-selection assay. Mol Cell Biol 16:4207–4214

Buchman AR, Berg P (1988) Comparison of intron-dependent and intron-independent gene expression. Mol Cell Biol 8:4395–405

Cáceres JF, Misteli T, Screaton GR, Spector DL, Krainer AR (1997) Role of the modular domains of SR proteins in subnuclear localization and alternative splicing specificity. J Cell Biol 138:225–238

Cáceres JF, Screaton GR, Krainer AR (1998) A specific subset of SR proteins shuttles continuously between the nucleus and the cytoplasm. Genes Dev 12:55–66

Chang DD, Sharp PA (1989) Regulation by HIV rev depends upon recognition of splice sites. Cell 59:789–795

Colgan DF, Manley JL (1997) Mechanism and regulation of mRNA polyadenylation. Genes Dev 11:2755–2766

Corbett AH, Silver PA (1997) Nucleocytoplasmic transport of macromolecules. Microbiol Mol Biol Rev 61:193–211

Coutavas E, Ren M, Oppenheim JD, D'Eustachio P, Rush MG (1993) Characterization of proteins that interact with the cell-cycle regulatory protein Ran/TC4. Nature 366:585–587

Cullen BR, Malim MH (1991) The HIV-1 rev protein – prototype of a novel class of eukaryotic post-transcriptional regulators. Trends Biochem Sci 16:346–350

Daneholt B (1997) A look at messenger RNP moving through the nuclear pore. Cell 88:585–588

Davis LI (1995) The nuclear pore complex. Annu Rev Biochem 64:865–896

Dingwall C, Laskey RA (1991) Nuclear targeting sequences – a consensus. Trends Biochem Sci 16:478–481

Doye V, Hurt E (1997) From nucleoporins to nuclear pore complexes. Curr. Opin. Cell Biol. 9:401–411

Dreyfuss G, Matunis MJ, Piñol-Roma S, Burd CG (1993) hnRNP proteins and the biogenesis of mRNA. Annu Rev Biochem 62:289–321

Dyhrmikkelsen H, Kjems J (1995) Inefficient spliceosome assembly and abnormal branch site selection in splicing of an HIV-1 transcript in vitro. J Biol Chem 270:24060–24066

Eckner R, Ellmeier W, Birnstiel ML (1991) Mature messenger RNA 3′ end formation stimulates RNA export from the nucleus. EMBO J 10:3513–3522

Englmeier L, Olivo JC, Mattaj IW (1999) Receptor-mediated substrate translocation through the nuclear pore complex without nucleotide triphosphate hydrolysis. Curr Biol 9:30–41

Ernst RK, Bray M, Rekosh D, Hammarskjold ML (1997) A structured retroviral RNA element that mediates nucleocytoplasmic export of intron-containing RNA. Mol Cell Biol 17:135–144

Fischer U, Meyer S, Teufel M, Heckel C, Lührmann R, Rautmann G (1994) Evidence that HIV-1 Rev directly promotes the nuclear export of unspliced RNA. EMBO J 13:4105–4112

Fischer U, Huber J, Boelens WC, Mattaj IW, Lührmann R (1995) The HIV-1 Rev activation domain is a nuclear export signal that accesses an export pathway used by specific cellular RNAs. Cell 82:475–483

Floer M, Blobel G (1999) Putative reaction intermediates in Crm1-mediated nuclear protein export. J Biol Chem 274:16279–16286

Fornerod M, Ohno M, Yoshida M, Mattaj IW (1997a) CRM1 is an export receptor for leucine-rich nuclear export signals. Cell 90:1051–1060

Fornerod M, van Deursen J, van Baal S, Reynolds A, Davis D, Murti KG, Fransen J, Grosveld G (1997b) The human homologue of yeast CRM1 is in a dynamic subcomplex with CAN/Nup214 and a novel nuclear pore component Nup88. EMBO J 16:807–816

Fridell RA, Fischer U, Lührmann R, Meyer BE, Meinkoth J, Malim MH, Cullen BR (1996) Amphibian transcription factor IIIA proteins contain a sequence element functionally equivalent to the nuclear export signal of human immunodeficiency virus type I Rev. Proc Natl Acad Sci USA 93:2936–2940

Fridell RA, Truant R, Thorne L, Benson RE, Cullen BR (1997) Nuclear import of hnRNP A1 is mediated by a novel cellular cofactor related to karyopherin-β. J Cell Sci 110:1325–1331

Fritz CC, Green MR (1996) HIV Rev uses a conserved cellular protein export pathway for the nucleocytoplasmic transport of viral RNAs. Curr Biol 6:848–854

Fu X-D (1995) The superfamily of arginine/serine-rich splicing factors. RNA 1:663–680

Ge H, Manley JL (1990) A protein factor, ASF, controls cell-specific alternative splicing of SV40 early pre-messenger RNA in vitro. Cell 62:25–34

Gorlach M, Burd CG, Dreyfuss G (1994) The mRNA poly(A)-binding protein: localization, abundance, and RNA-binding specificity. Exp Cell Res 211:400–407

Görlich D (1997) Nuclear protein import. Curr Opin Cell Biol 9:412–419

Görlich D, Mattaj IW (1996) Nucleocytoplasmic transport. Science 271:1513–1518

Görlich D, Kraft R, Koska S, Vogel F, Hartmann E, Laskey RA, Mattaj IW, Izaurralde E (1996) Importin provides a link between nuclear protein import and U snRNA export. Cell 87:21–32

Görlich D, Dabrowski M, Bischoff FR, Kutay U, Bork P, Hartmann E, Prehn S, Izaurralde E (1997) A novel class of RanGTP binding proteins. J Cell Biol 138:65–80

Gruss P, Lai C-J, Dhar R, Khoury G (1979) Splicing as a requirement for biogenesis of functional 16S mRNA of simian virus 40. Proc Natl Acad Sci USA 76:4317–4321

Grüter P, Tabernero C, von Kobbe C, Schmitt C, Saavedra C, Bachi A, Wilm M, Felber BK, Izaurralde E (1998) TAP, the human homolog of Mex67p, mediates CTE-dependent RNA export from the nucleus. Mol Cell 1:649–659

Hamer D, Leder P (1979) Splicing and the formation of stable RNA. Cell 18:1299–1302

Hamm J, Mattaj IW (1990) Monomethylated cap structures facilitate RNA export from the nucleus. Cell 63:109–118

Hattori K, Angel P, Beau MML, Karin M (1988) Structure and chromosomal localization of the functional intronless human JUN protooncogene. Proc Natl Acad Sci USA 85:9148–9152

Heaphy S, Finch JT, Gait MJ, Karn J, Singh M (1991) Human immunodeficiency virus type 1 regulator of virion expression, rev, forms nucleoprotein filaments after binding to a purine-rich 'bubble' located within the rev-responsive region of viral mRNAs. Proc Natl Acad Sci USA 88:7366–7370

Hentschel CC, Birnstiel MX (1981) The organization and expression of histone gene families. Cell 25:301–313

Hodge CA, Colot HV, Stafford P, Cole CN (1999) Rat8p/Dbp5p is a shuttling transport factor that interacts with Rat7p/Nup159p and gle1p and suppresses the mRNA export defect of xpo1-1 cells. EMBO J 18:5778–5788

Hopper AK, Traglia HM, Dunst RW (1990) The yeast RNA1 gene product necessary for RNA processing is located in the cytosol and is apparently excluded from the nucleus. J Cell Biol 111:309–321

Huang J, Liang TJ (1993) A novel hepatitis B virus (HBV) genetic element with Rev response element-like properties that is essential for expression of HBV gene products. Mol Cell Biol 13:7476–7486

Huang Y, Carmichael GG (1996a) Role of polyadenylation in nucleocytoplasmic transport of mRNA. Mol Cell Biol 16:1534–1542

Huang Y, Carmichael GG (1996b) A suboptimal 5′ splice site is a cis-acting determinant of nuclear export of polyomavirus late mRNAs. Mol Cell Biol 16:6046–6054

Huang Y, Carmichael GG (1997) The mouse histone H2a gene contains a small element that facilitates cytoplasmic accumulation of intronless gene transcripts and of unspliced HIV-1-related mRNAs. Proc Natl Acad Sci USA 94:10104–10109

Huang Z-M, Yen TSB (1995) Role of the hepatitis B virus posttranscriptional regulatory element in export of intronless transcripts. Mol Cell Biol 15:3864–3869

Huang Y, Wimler KM, Carmichael GG (1999) Intronless mRNA transport elements may affect multiple steps of pre- mRNA processing. EMBO J 18:1642–1652

Izaurralde E, Adam S (1998) Transport of macromolecules between the nucleus and the cytoplasm. RNA 4:351–364

Izaurralde E, Stepinski J, Darzynkiewicz E, Mattaj IW (1992) A cap binding protein that may mediate nuclear export of RNA polymerase II-transcribed RNAs. J Cell Biol 118:1287–1295

Izaurralde E, Lewis J, McGuigan C, Jankowska M, Darzynkiewicz E, Mattaj IW (1994) A nuclear cap binding protein complex involved in pre-mRNA splicing. Cell 78:657–668

Izaurralde E, Lewis J, Gamberi C, Jarmolowski A, McGuigan C, Mattaj IW (1995) A cap-binding protein complex mediating U snRNA export. Nature 376:709–712

Izaurralde E, Jarmolowski A, Beisel C, Mattaj IW, Dreyfuss G, Fischer U (1997) A role for the M9 transport signal of hnRNP A1 in mRNA nuclear export. J Cell Biol 137:27–35

Jarmolowski A, Boelens WC, Izaurralde E, Mattaj IW (1994) Nuclear export of different classes of RNA is mediated by specific factors. J Cell Biol 124:627–635

Jonsson JJ, Foresman MD, Wilson N, McIvor RS (1992) Intron requirement for expression of the human purine nucleoside phosphorylase gene. Nucleic Acids Res 20:3191–3198

Kataoka N, Ohno M, Kangawa K, Tokoro Y, Shimura Y (1994) Cloning of a complementary DNA encoding an 80 kDa nuclear cap binding protein. Nucleic Acids Res 22:3861–3865

Kedes LH (1979) Histone genes and histone messengers. Annu Rev Biochem 48:837–870

Koilka BK, Frielle T, Collins S, Yang-Feng T, Kobilka TS, Francke U, Lefkowitz RJ, Caron MG (1987) An intronless gene encoding a potential member of the family of receptors coupled to guanine nucleotide regulatory proteins. Nature 329:75–79

Koizumi J, Okamoto Y, Onogi H, Mayeda A, Krainer AR, Hagiwara M (1999) The subcellular localization of SF2/ASF is regulated by direct interaction with SR protein kinases (SRPKs). J Biol Chem 274:11125–11131

Krainer AR, Conway GC, Kozak D (1990) The essential pre-messenger RNA splicing factor-SF2 influences 5′ splice site selection by activating proximal sites. Cell 62:35–42

Kutay U, Izaurralde E, Bischoff FR, Mattaj IW, Görlich D (1997) Dominant-negative mutants of importin-beta block multiple pathways of import and export through the nuclear pore complex. EMBO J 16:1153–1163

Legrain P, Rosbash M (1989) Some cis- and trans-acting mutants for splicing target pre-mRNA to the cytoplasm. Cell 57:573–583

Liu X, Mertz J (1995) HnRNP L binds a cis-acting RNA sequence element that enables intron-independent gene expression. Genes Dev 9:1766–1780

Malim MH, Cullen BR (1991) HIV-1 structural gene expression requires the binding of multiple Rev monomers to the viral RRE: implications for HIV-1 latency. Cell 65:241–248

Manley JL, Tacke R (1996) SR proteins and splicing control. Genes Dev 10:1569–1579

Matunis EL, Matunis MJ, Dreyfuss G (1993) Association of individual hnRNP proteins and snRNPs with nascent transcripts. J Cell Biol 121:219–228

Mayeda A, Krainer AR (1992) Regulation of alternative pre-mRNA splicing by hnRNP A1 and splicing factor SF2. Cell 68:365–375

Mayeda A, Munroe SH, Cáceres JF, Krainer AR (1994) Function of conserved domains of hnRNP A1 and other hnRNP A/B proteins. EMBO J 13:5483–5495

Mehlin H, Daneholt B, Skoglund U (1995) Structural interaction between the nuclear pore complex and a specific translocating RNP particle. J Cell Biol 129:1205–1216

Melchior F, Weber K, Gerke V (1993) A functional homologue of the RNA1 gene produce in Schizosaccharomyces pombe: purification, biochemical characterization, and identification of a leucine-rich repeat motif. Mol Biol Cell 4:569–581

Melchior F, Guan T, Yokoyama N, Nishimoto T, Gerace L (1995) GTP hydrolysis by Ran occurs at the nuclear pore complex in an early step of protein import. J Cell Biol 131:571–581

Michael WM, Choi M, Dreyfuss G (1995a) A nuclear export signal in hnRNP A1: a signal-mediated, temperature- dependent nuclear protein export pathway. Cell 83:415–422

Michael WM, Siomi H, Choi M, Piñol-Roma S, Nakielny S, Liu Q, Dreyfuss G (1995b) Signal sequences that target nuclear import and nuclear export of pre- mRNA-binding proteins. Cold Spring Harbor Symp Quant Biol 60:663–668

Michael WM, Eder PS, Dreyfuss G (1997) The K nuclear shuttling domain: a novel signal for nuclear import and nuclear export in the hnRNP K protein. EMBO J 16:3587–3598

Murphy R, Wente SR (1996) An RNA-export mediator with an essential nuclear export signal. Nature 383:357–360

Nagata S, Mantei N, Weissmann C (1980) The structure of one of the eight or more distinct chromosomal genes for human interferon-alpha. Nature 287:401–408

Nakielny S, Dreyfuss G (1996) The hnRNP C proteins contain a nuclear retention sequence that can override nuclear export signals. J Cell Biol 134:1365–1373

Nakielny S, Dreyfuss G (1997) Nuclear export of proteins and RNAs. Curr Opin Cell Biol 9:420–429

Nakielny S, Dreyfuss G (1998) Import and export of the nuclear protein import receptor transportin by a mechanism independent of GTP hydrolysis. Curr Biol 8:89–95

Nakielny S, Fischer U, Michael WM, Dreyfuss G (1997) RNA transport. Annu Rev Neurosci 20:269–301

Nesic D, Cheng J, Maquat LE (1993) Sequences within the last intron function in RNA 3′-end formation in cultured cells. Mol Cell Biol 13:3359–3369

Neuberger MS, Williams GT (1988) The intron requirement for immunoglobulin gene expression is dependent upon the promoter. Nucleic Acids Res 16:6713–6724

Nigg EA (1997) Nucleocytoplasmic transport: signals, mechanisms and regulation. Nature 386:779–787

Ohno M, Kataoka N, Roberts MW (1990) A nuclear cap binding protein from HeLa cells. Nucleic Acids Res 18:6989–6995

Ohno M, Fornerod M, Mattaj IW (1998) Nucleocytoplasmic transport: the last 200 nm. Cell 92:327–336

Pasquinelli AE, Ernst RK, Lund E, Grimm C, Zapp ML, Rekosh D, Hammarskjold ML, Dahlberg JE (1997) The constitutive transport element (CTE) of Mason-Pfizer monkey virus (MPMV) accesses a cellular mRNA export pathway. EMBO J 16:7500–7510

Piñol-Roma S, Dreyfuss G (1992) Shuttling of pre-mRNA binding proteins between nucleus and cytoplasm. Nature 355:730–732

Piñol-Roma S, Dreyfuss G (1993) hnRNP proteins: localization and transport between the nucleus and the cytoplasm. Trends Cell Biol 3:151–155

Pollard VW, Michael WM, Nakielny S, Siomi MC, Wang F, Dreyfuss G (1996) A novel receptor-mediated nuclear protein import pathway. Cell 86:985–994

Richards SA, Lounsbury KM, Carey KL, Macara IG (1996) A nuclear export signal is essential for the cytosolic localization of the Ran binding protein, RanBP1. J Cell Biol 134:1157–1168

Rout MP, Wente SR (1994) Pores for thought: nuclear pore complex proteins. Trends Cell Biol 4:357–365

Ryu WS, Mertz JE (1989) Simian virus 40 late transcripts lacking excisable intervening sequences are defective in both stability in the nucleus and transport to the cytoplasm. J Virol 63:4386–4394

Saavedra C, Felber B, Izaurralde E (1997) The simian retrovirus-1 constitutive transport element, unlike the HIV-1 RRE, uses factors required for cellular mRNA export. Curr Biol 7:619–628

Santos-Rosa H, Moreno H, Simos G, Segref A, Fahrenkrog B, Pante N, Hurt E (1998) Nuclear mRNA export requires complex formation between Mex67p and Mtr2p at the nuclear pores. Mol Cell Biol 18:6826–6838

Schmitt C, von Kobbe C, Bachi A, Pante N, Rodrigues JP, Boscheron C, Rigaut G, Wilm M, Seraphin B, Carmo-Fonseca M, Izaurralde E (1999) Dbp5, a DEAD-box protein required for mRNA export, is recruited to the cytoplasmic fibrils of nuclear pore complex via a conserved interaction with CAN/Nup159p. EMBO J 18:4332–4347

Schwoebel ED, Talcott B, Cushman I, Moore MS (1998) Ran-dependent signal-mediated nuclear import does not require GTP hydrolysis by Ran. J Biol Chem 273:35170–35175

Segref A, Sharma K, Doye V, Hellwig A, Huber J, Lührmann R, Hurt E (1997) Mex67p, a novel factor for nuclear mRNA export, binds to both poly(A)+ RNA and nuclear pores. EMBO J 16:3256–3271

Shen EC, Henry MF, Weiss VH, Valentini SR, Silver PA, Lee MS (1998) Arginine methylation facilitates the nuclear export of hnRNP proteins. Genes Dev 12:679–691

Siomi H, Dreyfuss G (1995) A nuclear localization domain in the hnRNP A1 protein. J Cell Biol 129:551–560

Snay-Hodge CA, Colot HV, Goldstein AL, Cole CN (1998) Dbp5p/Rat8p is a yeast nuclear pore-associated DEAD-box protein essential for RNA export. EMBO J 17:2663–2676

Stade K, Ford CS, Guthrie C, Weis K (1997) Exportin 1 (Crm1p) is an essential nuclear export factor. Cell 90:1041–1050

Stutz F, Rosbash M (1998) Nuclear RNA export. Genes Dev 12:3303–3319

Stutz F, Izaurralde E, Mattaj IW, Rosbash M (1996) A role for nucleoporin FG repeat domains in export of human immunodeficiency virus type 1 Rev protein and RNA from the nucleus. Mol Cell Biol 16:7144–7150

Tabernero C, Zolotuknin AS, Velentin A, Pavlakis GN, Felber BK (1996) The posttranscriptional control element of the simian retrovirus type 1 forms an extensive RNA secondary structure necessary for its function. J Virol 70:5998–6011

Tseng SI, Weaver PL, Liu Y, Hitomi M, Tartakoff AM, Chang TH (1998) Dbp5p, a cytosolic RNA helicase, is required for poly(A)+ RNA export. EMBO J 17:2651–2662

Visa N, Alzhanova EA, Sun X, Kiseleva E, Bjorkroth B, Wurtz T, Daneholt B (1996a) A pre-mRNA-binding protein accompanies the RNA from the gene through the nuclear pores and into polysomes. Cell 84:253–264

Visa N, Izaurralde E, Ferreira J, Daneholt B, Mattaj IW (1996b) A nuclear cap-binding complex binds Balbiani ring pre-mRNA cotranscriptionally and accompanies the ribonucleoprotein particle during nuclear export. J Cell Biol 133:5–14

Wahle E (1991) A novel poly(A)-binding protein acts as a specificity factor in the second phase of messenger RNA polyadenylation. Cell 66:759–768

Wahle E, Kühn U (1997) The mechanism of 3' cleavage and polyadenylation of eukaryotic pre-mRNA. Prog Nucl Acids Res Mol Biol 57:41–71

Wu J, Matunis MJ, Kraemer D, Blobel G, Coutavas E (1995) Nup358, a cytoplasmically exposed nucleoporin with peptide repeats, Ran-GTP binding sites, zinc fingers, a cyclophilin A homologous domain, and a leucine-rich region. J Biol Chem 270:14209–14213

Yokoyama N, Hayashi N, Seki T, Pante N, Ohba T, Nishii K, Kuma K, Hayashida T, Miyata T et al (1995) A giant nucleopore protein that binds Ran/TC4. Nature 376:184–188

Zahler AM, Neugebauer KM, Lane WS, Roth MB (1993) Distinct functions of SR proteins in alternative pre-messenger RNA splicing. Science 260:219–222

Zapp ML, Hope TJ, Parslow TG, Green MR (1991) Oligomerization and RNA binding domains of the type 1 human immunodeficiency virus Rev protein: a dual function for an arginine-rich binding motif. Proc Natl Acad Sci USA 88:7734–7738

Zhao J, Hyman L, Moore C (1999) Formation of mRNA 3' ends in eukaryotes: mechanism, regulation, and interrelationships with other steps in mRNA synthesis. Microbiol Mol Biol Rev 63:405–445

Zolotukhin AS, Felber BK (1997) Mutations in the nuclear export signal of human Ran-binding protein RanBP1 block the Rev-mediated posttranscriptional regulation of human immunodeficiency virus type 1. J Biol Chem 272:11356–11360

Zolotukhin AS, Valentin A, Pavlakis GN, Felber BK (1994) Continuous propagation of RRE(-) and Rev(-)RRE(-) Human immunodeficiency virus type 1 molecular clones containing a cis-acting element of simian retrovirus type 1 in human peripheral blood lymphocytes. J Virol 68:7944–7952

RNA Localization in *Xenopus* Oocytes

Kinneret Rand and Joel Yisraeli[1]

1 Introduction

Because of the relatively large size of *Xenopus* oocytes (up to 1.4 mm in diameter by the end of stage VI), their ease of injection and manipulation, and their clearly defined animal-vegetal polarity, this system represents one of the most intensively studied systems for RNA localization. One of the first screens for localized RNAs was performed on *Xenopus* eggs and oocytes, and identified both animally and vegetally localized RNAs (Rebagliati et al. 1985). Since then, several different techniques have identified a number of additional RNAs, whose functions are quite varied (see King et al. 1999; Mowry and Cote 1999). In addition, *trans*-acting factors have been described that appear to play important roles in localizing RNAs in oocytes. The role cytoskeletal elements play in localizing various RNAs has also been examined. In short, the *Xenopus* oocyte system can be considered in many ways a model system for understanding how RNAs are intracellularly localized.

Throughout oogenesis, the animal-vegetal axis represents the axis of symmetry around which the oocyte appears to be cylindrically symmetric. RNA localization can be separated into at least three distinct pathways in these cells. RNAs arrive at the vegetal cortex via two different pathways. The early pathway, which occurs in stage I-II oocytes, is characterized by the localization of RNAs to a tight disk at the vegetal cortex, in a region where the germ plasm is located. The late pathway, occurring in stage III–IV oocytes, localizes RNAs to a broad arc that extends along the entire vegetal cortex. A third pathway of RNA localization involves the accumulation of a variety of RNAs at the animal pole of oocytes. In this chapter, we will review the literature on the localization of RNA along all three of these pathways in *Xenopus* oocytes. In particular, we will review the *cis*-acting elements that have been characterized for each pathway, the *trans*-acting factors, and the cytoskeletal elements involved. One of the striking conclusions that result from this analysis is that, despite the very specialized nature of the *Xenopus* oocyte, many of the same mechanisms, and, in some cases, the same or very similar proteins, are found to function in both oocytes and very disparate types of cells and organisms.

[1] Department of Anatomy and Cell Biology, Hebrew University – Hadassah Medical School, POB 12272, 91120 Jerusalem, Israel

Results and Problems in Cell Differentiation, Vol. 34
D. Richter (Ed.): Cell Polarity and Subcellular RNA Localization
© Springer-Verlag Berlin Heidelberg 2001

2 The Early Localization Pathway

The early pathway, which occurs in stage I–II oocytes, involves a gradual condensation and coalescence of these RNAs, concomitant, in time and location, with the formation of the mitochondrial cloud (MC; Kloc et al. 1996; see Fig. 1). The RNAs, initially distributed in several clusters around the germinal vesicle (GV), are targeted to the MC and take up characteristic positions within it. This structure, which has been termed the messenger transport organizer (METRO) then translocates to the vegetal cortex (Kloc and Etkin 1995). As the MC disperses, the comigrating RNAs are apparently anchored at the vegetal cortex in a fairly narrow arc. Translocation of RNAs in this pathway is not susceptible to drugs that depolymerize microtubules (MTs) or microfilaments (MFs), although maintenance of the RNAs at the cortex appears to require intact MFs (Kloc and Etkin 1995; Kloc et al. 1996).

2.1 RNAs That Undergo Localization Along the Early Pathway

The first RNA found to localize along the early pathway was Xlsirt, a family of non-coding RNAs found to contain repetitive sequences related to the chromosome X inactivation sequence, Xist (Kloc et al. 1993). Xcat-2 RNA was first identified in a screen for RNAs associated with the cytoskeleton and then found to associate with the MC (Mosquera et al. 1993; Forristall et al. 1995). Related to the *Drosophila* posterior pole gene *nanos*, Xcat2 remains associated with germ plasm throughout early *Xenopus* development. Xdazl and Xpat are two additional RNAs, localized along the early pathway, that are components of

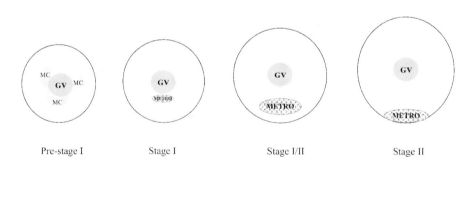

| Pre-stage I | Stage I | Stage I/II | Stage II |

RNAs

Fig. 1. Schematic version of the early RNA localization pathway. Note how RNAs localize to a single mitochondrial cloud (*MC*) in stage I oocytes, which has been selected to be the *METRO* (messenger transport organizer). Figure modeled after Kloc et al. (1996)

Table 1. RNAs localized to the vegetal pole via the early pathway

RNA	Function in development	Related to	Reference
DEADsouth (Xcat-3)		eIF4A	Forristall et al. (1995)
Xcat-2		Nanos	Mosquera et al. (1993)
Xdazl	Specification or	DAZ	Houston et al. (1998)
Xlsirts	development	Xsist RNA	Kloc et al. (1993)
Xpat	of germ cells	No homology	Hudson and Woodland (1998)
Xwnt-11		Wnts	Hudson and Woodland (1998); Ku and Melton (1993)
Fatvg		Adipophilin/ADRP	Chan et al. (1999)
Fingers (B7)	Unknown	Kruppel	Forristall et al. (1995)
XFACS (C10)		LA CoA Synthetase	Forristall et al. (1995)

germ plasm; depletion of maternal Xdazl RNA impairs the normal migration of primordial germ cells from the ventral endoderm (Houston and King 2000). MC association appears to be quite specific; Vg1 RNA, a late pathway member, is clearly excluded from the MC from stage I onwards (Kloc and Etkin 1995; Kloc et al. 1996, 1998). Inasmuch as germ plasm is set up early in oogenesis at the vegetal pole and then segregated in the early cleavages, many early pathway RNAs may play critical roles in forming, maintaining, and directing the germ cell lineage (see Table 1). Although a role for Xwnt11, another early pathway RNA, has not yet been definitely defined in *Xenopus*, wnt11 in zebra fish (Heisenberg et al. 2000), and the wingless pathway in *Xenopus* (Wallingford et al. 2000), appear to be crucial for correct cell polarity in cells migrating during gastrulation.

2.2 *cis*-Acting Elements

cis-Acting elements have been examined in two early pathway RNAs. In the case of Xlsirt-containing RNAs, it was shown that the repeat elements (~80 nt long) were necessary and sufficient for localization to the MC (Kloc et al. 1993; Kloc and Etkin 1998). In Xcat2, the signals have been examined more closely and appear to be more complex. Zhou and King (1996a) have identified a region at the beginning of the 3′ UTR which is necessary for localization to the MC upon injection of the message into stage I oocytes. (These investigators have also characterized a different *cis*-acting element that can localize injected Xcat2 RNAs to the vegetal cortex of stage IV oocytes in a manner similar to the late pathway RNAs; see below.) Recently, Kloc et al. (2000) have extended these findings and identified an additional *cis*-acting element, distal to the first one, in the 3′ UTR of Xcat2 which is necessary and sufficient for localizing transcripts to the germinal granules in early oocytes (termed the germinal granule localization element, GGLE). In particular, Xlsirt RNA, which does not associate with germinal granule material on its own (Kloc et al. 1998, 2000), is localized to this material when the *cis*-acting GGLE is attached to it (Kloc et al. 2000). In the absence of the mitochondrial localization element, however, the GGLE appears to be inactive (Kloc et al. 2000). These results suggest that *cis*-acting elements in the early pathway RNAs may direct localization to different intracellular targets, one associating the RNAs with the MC, and the other directing a subset of these RNAs to the germinal granules.

2.3 *trans*-Acting Factors and Cytoskeletal Elements

No *trans*-acting factors that interact specifically with the early pathway RNAs have yet been described. (Xlsirt and Xcat2 RNAs, however, are bound in vitro by a *trans*-acting factor, Vg1 RBP/Vera, thought to be involved in the late pathway; see below.) Our understanding of the cytoskeletal elements involved is only slightly better. Xcat2 RNA, initially identified in a screen for RNAs associated with the detergent-insoluble fraction (DIF) of oocytes, remains tightly associated with the DIF even after maturation (as opposed to Vg1 RNA, whose release upon maturation can be observed both biochemically and by in situ hybridization; Yisraeli et al. 1990; Forristall et al. 1995). Neither MTs nor MFs are required for METRO migration or early pathway RNA localization, however (Kloc and Etkin 1995; Kloc et al. 1996). Nevertheless, the anchoring of Xcat2, Xlsirt, and Xwnt11 RNAs at the vegetal cortex in stage III oocytes appears to be at least partially dependent on intact MFs (Kloc and Etkin 1995). The mechanism driving METRO localization to the vegetal cortex is still unclear, although an association of a subdomain of the endoplasmic reticulum (ER) with the migrating cloud has been reported (Kloc and Etkin 1998). γ-Tubulin, a component of the microtubule organizing center, also appears to be associated with the MC, perhaps as early as the 16-cell nest stage (Kloc and Etkin 1998). A spectrin-like protein is present in the mitochondrial cloud of stage I oocytes and remains associated with germ plasm throughout early cleavages (Kloc et al. 1998). Spectrin is a component of the spectrosome, required for the organization of the polarized MT-based RNA transport system in *Drosophila* egg chambers (Deng and Lin 1997).

3 The Late Pathway

The later pathway, which occurs in stage III–IV oocytes, involves a gradual disappearance of the localized RNAs from the animal hemisphere with a concomitant accumulation in the vegetal hemisphere and around the GV in late stage III oocytes (Melton 1987; Yisraeli and Melton 1988; see Fig. 2). By the beginning of stage IV, these RNAs appear to be tightly anchored in a thin shell along a broad arc that extends essentially throughout the vegetal cortex. A tran-

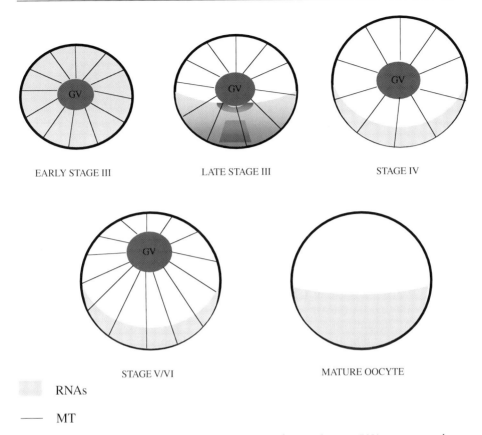

EARLY STAGE III   LATE STAGE III   STAGE IV

STAGE V/VI   MATURE OOCYTE

RNAs

—— MT

Fig. 2. Schematic version of the late RNA localization pathway, using Vg1 RNA as an example. The orientation of the MTs (microtubules) during these stages of oogenesis is likely to be of mixed polarity (Gard et al. 1995). Note that in the mature oocyte, GV (germinal vesicle) breakdown has occurred, and the cytoskeleton is undergoing reorganization

sient wedge-like distribution of at least some of the localizing RNA, extending from the vegetal aspect of the GV to the cortex, has also been described (Kloc and Etkin 1995). As opposed to the early pathway, translocation of RNAs along the late pathway is disrupted by MT depolymerization; maintenance of the RNA at the cortex requires intact MFs (Yisraeli et al. 1990).

3.1 RNAs That Undergo Localization Along the Late Pathway

Vg1 RNA was isolated in the first screen for RNAs localized along the animal-vegetal axis in *Xenopus* oocytes (Rebagliati et al. 1985). A member of the TGFβ superfamily, Vg1 has been implicated in both endoderm development and mesoderm induction (Weeks and Melton 1987; Thomsen and Melton 1993; Joseph and Melton 1998). Two other RNAs implicated in germ layer formation

Table 2. RNAs localized to the vegetal pole using the late pathway

RNA	Function in development	Related to	Reference
β-TrCP-2	Cell cycle	β-Transducin	Hudson et al. (1996)
β-TrCP-3	Cell cycle	β-Transducin	Hudson et al. (1996)
Fatvg	Unknown	Adipophilin/ADRP	Chan et al. (1999)
VegT	Primary germ layers	Brachyury	Horb and Thomsen (1997); Lustig et al. (1996); Stennard et al. (1996); Zhang and King (1996)
Vg1	Meso/Endo, R/L axis	TGFβ superfamily	Rebagliati et al. (1985)
Xcat-4	Unknown	Unknown	Forristall et al. (1995)
X-Bic-c	Endoderm induction	Bicaudal-C	Wessely and De Robertis (2000)

have also been found to localize along the late pathway. VegT is a T-box transcription factor involved in endoderm and mesoderm patterning; VegT RNA is localized temporally and spatially in oocytes in a manner very similar to Vg1 RNA, although its release from the vegetal cortex appears to precede that of Vg1 RNA (Stennard et al. 1996; Zhang and King 1996). Xbic-C has very recently been cloned and appears to act as a translational repressor that induces endoderm formation (Wessely and De Robertis 2000). Its RNA is localized in a shell along the vegetal cortex in stage IV oocytes, although it also appears to be distributed in a graded fashion extending up to the animal hemisphere. Other late pathway localized RNAs are shown in Table 2. Several RNAs (Xcat-2, Xlsirts, Xpat) localize normally along the early pathway but will also localize like Vg1 RNA when injected into stage III oocytes (Kloc et al. 1993; Zhang and King 1996; Hudson and Woodland 1998). In addition, rat tau RNA, which localizes to the proximal hillock of rat axons, also localizes to the vegetal cortex when injected into *Xenopus* stage III oocytes, in a fashion that precisely mimics the localization of Vg1 RNA (Litman et al. 1996).

A recently described vegetally localized RNA isolated from a cDNA library enriched in MC mRNAs, termed fatvg, appears to blur the distinction, to some extent, between early and late pathways (Chan et al. 1999). Although fatvg RNA, as opposed to all of the other early pathway RNAs, does not associate with the MC in stage I oocytes, it gets localized to the center of the cloud in stage II oocytes. From stage III oocytes onward, its localization mirrors that of Vg1 and the other late pathway RNAs, accumulating along the vegetal aspect of the GV and then spreading out along a broad arc at the vegetal cortex. Thus, the localization of fatvg RNA appears to contain elements of both the early and late pathways.

3.2 *cis*-Acting Elements

The most extensively examined *cis*-acting element in RNAs localized in *Xenopus* oocytes is that of Vg1 RNA. Both in vivo and in vitro analyses have

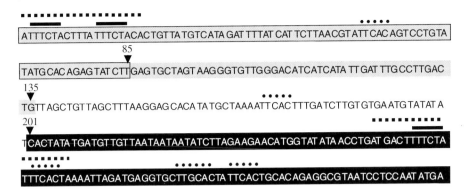

ATTTCTACTTTA TTTCTACACTGTTA TGTCATA GAT TTTAT CAT TCTTAACGTATTCAC AGTCCTGTA
85

TATGCAC AGAG TAT CTT GAGTGCTAGT AAGGGTGTTGGGACATCATCATA TT GAT TTGCCTTGAC
135

TGTTAGCTGTTAGCTTTAAGGAG CACA TA TGCTAAAATTCAC TTTGATCTTGTGTGAATGTATAT A
201

CACTATA TGATGT TGTTAATAATAATATCTTAGAAGAACATGGTAT ATA ACCTGAT GACTTTTCTA

TTTCACTAAAATTAGATGAGGTGCTTGCACTA TTCACTGCAC AGAGGCGTAATCCTCCAAT ATGA

AAAT AAAAAAATA TAT AAT AGTCCATAT GACTTG

——— VM1 (Gautreau et al., 1997)

▪▪▪▪▪ Binding and localization (Havin et al., 1998)

••••• Binding and localization (Deshler et al., 1997)

Fig. 3. The nucleotide sequence of the Vg1 VLE. The nucleotide sequence of the Vg1 VLE is shown, with the elements described in the text indicated above the sequence. The 5′ subelement, as mentioned in the text, consists of either sequences 1–85 (Cote et al. 1999) or 1–135 (*shaded in gray*, with or without a *frame*, respectively; Gautreau et al. 1997). The 3′ subelement, as mentioned in the text, consists of sequences from 201–340 (*shaded in black*). (Gautreau et al. 1997)

been performed by several groups (Mowry and Melton 1992; Schwartz et al. 1992; Mowry 1996; Deshler et al. 1997, 1998; Gautreau et al. 1997; Havin et al. 1998; Cote et al. 1999). The results are summarized in Fig. 3. All of these groups have made use of an oocyte culture system that allows one to incubate stage III–IV oocytes in medium supplemented with vitellogenin-containing frog serum (Keem et al. 1979; Wallace et al. 1980). Both endogenous and injected synthetic Vg1 mRNA localize in cultured oocytes in 4–6 days in a manner indistinguishable from Vg1 RNA localized in vivo (Yisraeli and Melton 1988). A 340-nt element in the 3′ UTR that is both necessary and sufficient for Vg1 RNA localization in oocytes (termed the VLE) was identified by ligating Vg1 RNA sequences to a β-globin reporter gene and assaying localization in vivo (Mowry and Melton 1992). A more detailed mapping of the fine structure of this element has suggested that its composition is complex. Deletion analyses and linker scans spanning the entire VLE suggest that single, short deletions or replacements (of between 5–20 nt) tend to have, at best, marginal effects on localization (Gautreau et al. 1997; Deshler et al. 1998; Havin et al. 1998). Larger, single deletions near either end of the VLE (of 40–65 nt) or a double replacement of two 20-nt fragments with vector sequences of the same length can successfully abolish localization. These results suggest that the localization signal is probably a compound element consisting of two subelements at or near the ends of the VLE. Furthermore, relatively large deletions between these subelements have no effect on localization. The precise nature of the *cis*-acting sequences that are active in these subelements is still unclear. Deletion of three interspersed 5- to 6-nt repeat sequences (UUCAC or UUGCAC) in the downstream subelement nearly abolishes localization (Deshler et al. 1998). Only the upstream, and not the downstream, subelement, however, is capable of directing localization when tandemly repeated. A 6-nt sequence (UUUCUA), termed VM1 (Gautreau et al. 1997), is present in two copies in the first subelement and in a single copy in the second subelement. Mutating the four VM1 sequences in a tandemly repeated upstream subelement construct disrupts localization, suggesting that VM1 is a necessary component of the *cis*-acting signal. Tandemly repeated elements, however, do not necessarily fold or behave precisely like the larger sequence from which they are derived. It is important to assay these mutations in the context of the intact VLE in order to ensure that these sequences play a role in vivo, and so far this has not been done.

Two other late pathway localization elements have been analyzed. As mentioned above, although endogenous Xcat-2 RNA appears to be localized along the early pathway, Xcat-2 RNA injected into stage III oocytes is localized to the vegetal cortex in a manner that appears to be identical to Vg1 RNA. A bipartite *cis*-acting element that is both necessary and sufficient for this late pathway localization has been mapped (Zhou and King 1996b). Interestingly, a VM1 sequence is present in the more proximal part of this *cis*-acting element (JKY, pers. observ.). It should be noted that the mitochondrial localization element in Xcat-2 contains the proximal part of the late pathway *cis*-acting element (Zhou and King 1996a), while the germinal granule localization element contains the distal part of the element (Kloc et al. 2000). The functional relationship between these various localization elements remains to be determined. Fatvg RNA, although also demonstrating some early pathway localization behavior, localizes in large measure along the late pathway (Chan et al. 1999). These researchers have analyzed the late pathway localization element in fatvg RNA and identified a 25-nt fragment that appears to be sufficient for localization (a second downstream element apparently able to act independently was also observed). None of the repetitive motifs that have been proposed to play a role in late pathway localization are either present or appear to be important for this element. This novel localization element may reflect the unusual nature of the RNA.

3.3 *trans*-Acting Factors

Although there is an array of RNAs that localize to different regions of the oocyte at different stages of oogenesis, work in the *Xenopus* field on *trans*-acting factors has focused so far on those that interact with Vg1 RNA. Vg1 RBP was identified as a 69-kDa protein in S100 extracts that shows high affinity binding to the Vg1 VLE in UV cross-linking assays (Schwartz et al. 1992). Vg1

RBP is enriched in oocyte MT preparations and pellets with semi-purified, polymerized tubulin added to oocyte protein extracts. In addition, Vg1 RBP helps direct the specific association of Vg1 RNA to reconstituted MTs in vitro (Elisha et al. 1995). A similarly sized protein (75 kDa) was identified in oocyte low-speed supernatants that also specifically interacted with the Vg1 VLE; this protein was termed Vera because of its ability to bind Vg1 RNA and its apparent endoplasmic reticulum association (Deshler et al. 1997). Both Vg1 RBP and Vera have been purified and cloned using methods that included Vg1 VLE RNA affinity chromatography, and these proteins turn out to be 100% identical in sequence (Deshler et al. 1998; Havin et al. 1998). It is currently unclear whether the difference in the observed size of the proteins is a result of post-translational modifications or slight variations in electrophoresis conditions. Vg1 RBP/Vera consists almost entirely of potential RNA binding domains: two RRM motifs, an RGG box, and four KH domains. The sequences mediating MT and/or ER associations have not yet been determined, although both RRM and KH domains can mediate protein–protein interactions (Chen et al. 1997; Samuels et al. 1998). It is also unknown at the moment which domains are responsible for the specific interaction of Vg1 RBP/Vera with Vg1 VLE. Intriguingly, there are putative nuclear localization and export sequences present in the protein, suggesting that Vg1 RBP/Vera may spend at least some of its life in the nucleus. In fact, an allele, 97% identical at the protein level, was also isolated during the cloning of Vg1 RBP and is identical to a previously described factor, B3 that binds to the promoter of TFIIIA and appears to participate in its transcriptional activation in early *Xenopus* oocytes (Pfaff and Taylor 1992). Antibodies raised to Vg1 RBP/Vera and to B3 indicate that in early oocytes, these proteins are present in the nucleus as well as the cytoplasm (Zhang et al. 1999). Vg1 RBP/Vera and Vg1 RNA distributions are coincident throughout oogenesis, initially homogeneous and then localizing vegetally with the RNA as it translocates and gets anchored at the cortex.

UV cross-linking has revealed a number of additional proteins that interact with Vg1 VLE as well (Mowry 1996). One of these, a 60-kDa protein termed VgRBP60, has recently been purified and cloned, based on its ability to bind specifically to the predominantly pyrimidine sequence, VM1 (Cote et al. 1999). VgRBP60 contains four RRM domains and is highly related to the human hnRNP proteins, hnRNP I and PTB (polypyrimidine tract binding protein). In the case of the human hnRNP I protein, RRM3 and RRM4 are thought to be the important RNA binding domains, with the N-terminal half of the protein believed to be involved in protein–protein interactions (Perez et al. 1997; Oh et al. 1998). As does Vg1 RBP/Vera, Vg1RBP60 colocalizes with Vg1 RNA to the vegetal cortex (Cote et al. 1999). A nuclear localization sequence (NLS), however, has also been detected in VgRBP60, a sequence conserved in hnRNP I. The ability of the hnRNP family proteins to shuttle between the nucleus and cytoplasm may be important for the role they play in RNA localization. Mutations in the gene *squid*, which encodes an hnRNP protein, disrupt grk mRNA localization in *Drosophila* oocytes (Norvell et al. 1999). In addition, hnRNP A2 has been shown to associate with a *cis*-acting element in MBP mRNA that is necessary and sufficient for its transport into peripheral processes in oligo-dendrocytes (Hoek et al. 1998).

How do these *trans*-acting factors interact specifically with the *cis*-acting VLE and what are their roles in localization? In the case of VgRBP60, mutagenesis evidence shows that the protein recognizes the VM1 sequences in the VLE, certainly the two copies in the 5′ subelement, and perhaps the third copy in the 3′ subelement (see Fig. 3). This correlates very nicely with the ability of constructs containing either wild type (wt) 5′ and 3′ subelements or a tandemly duplicated 5′ subelement to be localized, but not a mutated, tandemly duplicated 5′ subelement (Gautreau et al. 1997). In the case of Vg1 RBP/Vera, the story is less clear, with the answer somewhat dependent on the method used. A linker scan analysis, in which 20-nt-long sequences were progressively replaced by a 20-nt-long vector sequence throughout the length of the VLE, revealed that only two replacements affected UV cross-linking to Vg1 RBP/Vera (Havin et al. 1998). By preserving the total length of the VLE, while making limited sequence changes, this approach may minimize the effects of the changes on the overall secondary and tertiary structure of the RNA. Strikingly, one of these replacements removed both VM1 sequences from the 5′ subelement, and the other replacement removed the VM1 sequence from the 3′ subelement (see Fig. 3). Vg1 RBP/Vera was found to bind efficiently to the 5′ subelement only when it was a minimum length of 135 nt (Havin et al. 1998; Z. Elisha, unpubl. observ.). There was a good correlation between the ability of these VLE constructs to bind the protein and their ability to localize upon injection into stage III oocytes. A good correlation between Vg1 RBP/Vera binding and localization was also found when the sequences UUCAC or UUGCAC, which appear in five copies in the VLE were either deleted entirely, or in part (Deshler et al. 1997, 1998). Only one of these repeats is contained in the 20-nt sequences mentioned above. Although a small RNA probe containing four copies of this sequence and consisting of only 23 nt can cross-link Vg1 RBP/Vera, it is hard to deduce the specificity of the protein on the endogenous RNA from these results, with such an artificial substrate. Recent RNA footprinting results in our laboratory confirm that Vg1 RBP/Vera binds, at least in vitro, to the regions including and surrounding the VM1 repeats (K.R. and J.K.Y., unpubl. observ.).

Although these proteins bind to the same region of the VLE, it is unclear whether they interact, and how they contribute to the localization of Vg1 RNA. According to Cote et al. (1999), the wt tandemly repeated 5′ subelement gets localized and appears to bind well to VgRBP60 but poorly to Vg1 RBP/Vera. Furthermore, mutations in VM1 (in the context of the tandemly repeated 5′ element) prevent localization and severely reduce binding to VgRBP60, but have little effect on (the minimal) Vg1 RBP/Vera binding. On the basis of these results, Cote et al. argue that Vg1 RBP/Vera does not appear to interact with VM1. This argument seems premature, however. The size of the tandem repeat used in these constructs is 85 nt, which, as mentioned above (Z. Elisha, unpubl.

observ.), and as Cote et al. themselves show, does not support efficient Vg1 RBP/Vera binding. In fact, the longer, 135-nt tandem, repeat localizes more efficiently (Gautreau et al. 1997). It therefore appears that Vg1 RBP/Vera is unlikely to play a role in the (partial) localization of the tandemly repeated short 5' subelement. With respect to Vg1 RBP/Vera interaction with VM1 sequences, however, the jury is still out. Given the potential for artifactual folding of RNA and the artificial nature of short, repeated constructs, only experiments performed in the context of the full-length VLE are likely to allow analysis of the full complement of the factors involved.

3.4 Cytoskeletal Elements

The cytoskeletal elements involved in late pathway localization have been best characterized for Vg1 RNA localization. Using MT and MF inhibitors with the oocyte culture system described above, it was possible to identify at least two distinct steps in the localization process (Yisraeli et al. 1990). The first, susceptible to the MT inhibitors nocodozole, colchicine, or tubulozole C, is the accumulation of Vg1 mRNA in the vegetal hemisphere, and along the vegetal aspect of the GV, with a concomitant disappearance of mRNA from the animal hemisphere. The second step, susceptible to the MF inhibitor cytochalasin B, involves the anchoring of mRNA along the vegetal cortex. Exogenous mRNAs such as Xcat2 and tau, observed to undergo localization along the late pathway when injected into stage III oocytes, demonstrate similar susceptibilities to cytoskeletal inhibitors (Litman et al. 1996; Zhou and King 1996b). The involvement of MTs in transport of the RNA and MFs in its anchoring also corresponds with the location of these elements in stage III oocytes: MTs are radially arrayed running from the GV to the cortex, and MFs are present, although not exclusively, along the cortex (Franke et al. 1976; Yisraeli et al. 1990; Gard 1991; Roeder and Gard 1994). Vg1 mRNA has also been reported to colocalize in stage III oocytes with subfractions of the ER (Deshler et al. 1997; Kloc and Etkin 1998), although the functional significance of this association to the localization process has yet to be demonstrated. Depletion of the non-coding, repetitive, early pathway RNA Xlsirt from stage IV oocytes has been shown to release Vg1 RNA from its tight, cortical localization (Kloc and Etkin 1994). It is unclear, however, given the large amounts of antisense phosphothiolate oligonucleotides required for Xlsirt destruction in this experiment, whether the cortical cytoskeleton remained intact.

Vg1 mRNA undergoes a further change in its intracellular distribution at the end of oogenesis, when it is released from its tight cortical localization during maturation and fills up most of the vegetal hemisphere in unfertilized eggs (Weeks and Melton 1987). Cytochalasin B treatment of stage VI oocytes can essentially mimic this response, further substantiating a role for intact MFs in some aspect of the anchoring of Vg1 mRNA at the cortex; MT inhibitors have no affect on this localization (Yisraeli et al. 1990). It remains unclear which cytoskeletal elements are responsible for retaining Vg1 mRNA at the cortex.

MFs appear to play a crucial role in maintaining the integrity of the cortex. Cytochalasin B treatment of stage IV oocytes causes a release of both γ-tubulin and cytokeratins from the cortex, and the redistributed γ-tubulin precisely mirrors the redistribution of Vg1 mRNA following the same treatment (Gard 1994; Gard et al. 1997). It is important to note, however, that Vg1 mRNA was found to be enriched in a detergent-insoluble fraction of oocytes rich in cyto-keratins but depleted of tubulin and actin (Pondel and King 1988). Although Klymkowsky and colleagues (1991) have noted that cytokeratin severing occurs long after Vg1 mRNA release, more recent observations have observed a fine cortical cytokeratin meshwork with F-actin dependent links to the cortex (Gard et al. 1997). This layer, which demonstrates an animal-vegetal polarity, could form the basis for Vg1 mRNA anchoring (King et al. 1999). EM in situ hybridization combined with the use of specific antibodies to various cytoskeletal proteins should help pinpoint the elements involved.

4 The Animal Localization Pathway

A number of different RNAs have been found to be localized to the animal hemisphere of late stage oocytes, many of them thought to be involved in tran-scription or signal transduction (see Table 3). Animal localization has been a more difficult phenomenon to study than vegetal localization, given the fact that the majority of the yolk-free cytoplasm is located in the animal hemi-sphere and that many RNAs, when injected into the vegetal hemisphere, will passively diffuse and accumulate in the animal hemisphere (Drummond et al. 1985). Thus, there are not yet any *cis*-acting elements reported to direct animal localization. Nevertheless, there appear to be at least four different patterns of animal localization:

1. Localization throughout the animal hemisphere – e.g., An1, 2, 3 (Rebagliati et al. 1985), and XGβ1 (Devic et al. 1996);
2. Localization at the animal pole cortex – PABP (Schroeder and Yost 1996), Vg1 RBP (Zhang et al. 1999);
3. Localization perinuclearly – fibronectin (Oberman and Yisraeli 1995);
4. Localization in the GV and along the animal cortex – α-spectrin (Carotenuto et al. 2000).

It is currently unclear whether these patterns represent different pathways and/or different mechanisms for animal localization. In particular, it remains an open question as to whether animal localization is an active or passive process, and which cytoskeletal elements may be involved. In the case of Vg1 RBP mRNA localization to the animal hemisphere, the total level of Vg1 RBP mRNA appears to drop dramatically as the RNA becomes localized. This has raised the interesting possibility that specific degradation of the mRNA, perhaps coupled with protection of the mRNA at the animal cortex, may be responsible for its ultimate distribution. Such a mechanism for RNA localiza-tion has been demonstrated for hsp83 mRNA localization in *Drosophila*

Table 3. RNAs localized to the animal pole

RNA	Function in development	Related to	Reference
α-Spectrin	Formation of specific domains during oogenesis	Spectrin	Carotenuto et al. (2000)
βTrCP	Cell cycle	β-Transducin	Hudson et al. (1996)
An-1	Required beyond gastrulation	Unknown	Rebagliati et al. (1985)
An-2	Required for energy production	α F0/F1 ATPase	Rebagliati et al. (1985)
An-3	RNA processing in oocytes, RNA processing in embryos and adults	Mouse PL10	Rebagliati et al. (1985)
An-4	Brain development	Human KIAA0095	Hudson et al. (1996)
Oct-60	Unknown	Transcription factor	Weeks et al. (1995)
PABP	Translational control	Poly(A)-binding protein	Schroeder and Yost (1996)
Xl-21	Unknown	Transcription factor	Weeks et al. (1995)
Xlan-4	Specification of neural cell types, neurogenesis	Unknown	Reddy et al. (1992)
Fibronectin	Extracellular matrix protein	Fibronectin	Oberman and Yisraeli (1995)
Vg1 RBP/Vera	RNA binding protein involved in intracellular RNA localization	ZBP-1	Zhang et al. (1999)
XGβ1	Signal transduction	β1 Subunit of heteromeric GTP-binding protein	Devic et al. (1996)
XlcaaX (Xlgv7)	Osmoregulation via association with ion transporter or pump?	Unknown	Weeks et al. (1995)

oocytes, and a conserved degradation mechanism exists in *Xenopus* oocytes as well (Bashirullah et al. 1999).

5 Concluding Remarks

The number of localized RNAs in *Xenopus* oocytes has steadily risen since the first cDNAs were identified in the mid- 1980s. Progress has been made in identifying some *cis*-acting elements that help direct RNAs to particular intracellular locations, although it is still impossible to search a library for other RNAs that contain these elements, or even to identify these elements by inspection of known RNAs. The role of *trans*-acting factors in localization is just beginning to be understood. As has recently been described in *Drosophila* oocytes (Schnorrer et al. 2000), it is likely that at least some of these factors will interact with molecular motors or mediate interaction with a complex of proteins which contain such motors. Of particular interest will be how the growing body of knowledge about localized RNAs will help us understand more about how the polarity in oocytes is generated and interpreted.

Acknowledgements. This work has been supported in part by grants from the Israel Academy of Arts and Sciences and from the US-Israel Binational Science Foundation (J.K.Y.).

References

Bashirullah A, Halsell SR, Cooperstock RL, Kloc M, Karaiskakis A, Fisher WW, Fu W, Hamilton JK, Etkin LD, Lipshitz HD (1999) Joint action of two RNA degradation pathways controls the timing of maternal transcript elimination at the midblastula transition in *Drosophila melanogaster*. EMBO J 18:2610–2620

Carotenuto R, Vaccaro MC, Capriglione T, Petrucci TC, Campanella C (2000) α-Spectrin has a stage-specific asymmetrical localization during *Xenopus* oogenesis. Mol Reprod Dev 55: 229–239

Chan AP, Kloc M, Etkin LD (1999) fatvg encodes a new localized RNA that uses a 25-nucleotide element (FVLE1) to localize to the vegetal cortex of *Xenopus* oocytes. Development 126: 4943–4953

Chen T, Damaj BB, Herrera C, Lasko P, Richard S (1997) Self-association of the single-KH-domain family members Sam68, GRP33, GLD-1, and Qk1: role of the KH domain. Mol Cell Biol 17:5707–5718

Cote CA, Gautreau D, Denegre JM, Kress TL, Terry NA and Mowry KL (1999) A *Xenopus* protein related to hnRNP I has a role in cytoplasmic RNA localization. Mol Cell 4:431–437

Deng W, Lin H (1997) Spectrosomes and fusomes anchor mitotic spindles during asymmetric germ cell divisions and facilitate the formation of a polarized microtubule array for oocyte specification in *Drosophila*. Dev Biol 189:79–94

Deshler JO, Highett MI, Schnapp BJ (1997) Localization of *Xenopus* Vg1 mRNA by Vera protein and the endoplasmic reticulum. Science 276:1128–1131

Deshler JO, Highett MI, Abramson T, Schnapp BJ (1998) A highly conserved RNA-binding protein for cytoplasmic mRNA localization in vertebrates. Curr Biol 8:489–496

Devic E, Paquereau L, Rizzoti K, Monier A, Knibiehler B, Audigier Y (1996) The mRNA encoding a beta subunit of heterotrimeric GTP-binding proteins is localized to the animal pole of *Xenopus laevis* oocyte and embryos. Mech Dev 59:141–151

Drummond D, McCrae M, Colman A (1985) Stability and movement of mRNAs and their encoded proteins in *Xenopus* oocytes. J Cell Biol 100:1148–1156

Elisha Z, Havin L, Ringel I, Yisraeli JK (1995) Vg1 RNA binding protein mediates the association of Vg1 RNA with microtubules in *Xenopus* oocytes. EMBO J 14:5109–5114

Forristall C, Pondel M, Chen L, King ML (1995) Patterns of localization and cytoskeletal association of two vegetally localized RNAs, Vg1 and Xcat-2. Development 121:201–208

Franke WW, Rathke PC, Seib E, Trendelenburg MF, Osborn M, Weber K (1976) Distribution and mode of arrangement of microfilamentous structures and actin in the cortex of the amphibian oocyte. Cytobiologie 14:111–130

Gard DL (1991) Organization, nucleation, and acetylation of microtubules in *Xenopus laevis* oocytes: a study by confocal immunofluorescence microscopy. Dev Biol 143:346–362

Gard DL (1994) γ-Tubulin is asymmetrically distributed in the cortex of *Xenopus* oocytes. Dev Biol 161:131–140

Gard DL, Affleck D, Error BM (1995) Microtubule organization, acetylation, and nucleation in *Xenopus laevis* oocytes. II. A developmental transition in microtubule organization during early diplotene. Dev Biol 168:189–201

Gard DL, Cha BJ, King E (1997) The organization and animal-vegetal asymmetry of cytokeratin filaments in stage VI *Xenopus* oocytes is dependent upon F-actin and microtubules. Dev Biol 184:95–114

Gautreau D, Cote CA, Mowry KL (1997) Two copies of a subelement from the Vg1 RNA localization sequence are sufficient to direct vegetal localization in *Xenopus* oocytes. Development 124:5013–5020

Havin L, Git A, Elisha Z, Oberman F, Yaniv K, Schwartz SP, Standart N, Yisraeli JK (1998) RNA-binding protein conserved in both microtubule- and microfilament-based RNA localization. Genes Dev 12:1593–1598

Heisenberg CP, Tada M, Rauch GJ, Saude L, Concha ML, Geisler R, Stemple DL, Smith JC, Wilson SW (2000) Silberblick/Wnt11 mediates convergent extension movements during zebrafish gastrulation. Nature 405:76–81

Hoek KS, Kidd GJ, Carson JH, Smith R (1998) hnRNP A2 selectively binds the cytoplasmic transport sequence of myelin basic protein mRNA. Biochemistry 37:7021–7029

Horb ME, Thomsen GH (1997) A vegetally localized T-box transcription factor in *Xenopus* eggs specifies mesoderm and endoderm and is essential for embryonic mesoderm formation. Development 124:1689–1698

Houston DW, King ML (2000) A critical role for Xdazl, a germ plasm-localized RNA, in the differentiation of primordial germ cells in *Xenopus*. Development 127:447–456

Houston DW, Zhang J, Maines JZ, Wasserman SA, King ML (1998) A *Xenopus* DAZ-like gene encodes an RNA component of germ plasm and is a functional homologue of *Drosophila* boule. Development 125:171–180

Hudson C, Woodland HR (1998) Xpat, a gene expressed specifically in germ plasm and primordial germ cells of *Xenopus laevis*. Mech Dev 73:159–168

Hudson JW, Alarcon VB, Elinson RP (1996) Identification of new localized RNAs in the *Xenopus* oocyte by differential display PCR. Dev Genet 19:190–198

Joseph EM, Melton DA (1998) Mutant Vg1 ligands disrupt endoderm and mesoderm formation in *Xenopus* embryos. Development 125:2677–2685

Keem K, Smith LD, Wallace RA, Wolf D (1979) Growth rate of oocytes in laboratory-maintained *Xenopus laevis*. Gamete Res 2:125–135

King ML, Zhou Y, Bubunenko M (1999) Polarizing genetic information in the egg: RNA localization in the frog oocyte. Bioessays 21:546–557

Kloc M, Etkin LD (1994) Delocalization of Vg1 mRNA from the vegetal cortex in *Xenopus* oocytes after destruction of Xlsirt RNA. Science 265:1101–1103

Kloc M, Etkin LD (1995) Two distinct pathways for the localization of RNAs at the vegetal cortex in *Xenopus* oocytes. Development 121:287–297

Kloc M, Etkin LD (1998) Apparent continuity between the messenger transport organizer and late RNA localization pathways during oogenesis in *Xenopus*. Mech Dev 73:95–106

Kloc M, Spohr G, Etkin LD (1993) Translocation of repetitive RNA sequences with the germ plasm in *Xenopus* oocytes. Science 262:1712–1714

Kloc M, Larabell C, Etkin LD (1996) Elaboration of the messenger transport organizer pathway for localization of RNA to the vegetal cortex of *Xenopus* oocytes. Dev Biol 180:119–130

Kloc M, Larabell C, Chan APY, Etkin LD (1998) Contribution of METRO pathway localized molecules to the organization of the germ cell lineage. Mech Dev 75:81–93

Kloc M, Bilinski S, Pui-Yee Chan A, Etkin LD (2000) The targeting of Xcat2 mRNA to the germinal granules depends on a *cis*-acting germinal granule localization element within the 3'UTR. Dev Biol 217:221–229

Klymkowsky MW, Maynell LA, Nislow C (1991) Cytokeratin phosphorylation, cytokeratin filament severing and the solubilization of the maternal mRNA Vg1. J Cell Biol 114:787–797

Ku M, Melton DA (1993) Xwnt-11: a maternally expressed *Xenopus* wnt gene. Development 119:1161–1173

Litman P, Behar L, Elisha Z, Yisraeli JK, Ginzburg I (1996) Exogenous tau RNA is localized in oocytes: possible evidence for evolutionary conservation of localization mechanisms. Dev Biol 176:86–94

Lustig KD, Kroll KL, Sun EE, Kirschner MW (1996) Expression cloning of a *Xenopus* T-related gene (Xombi) involved in mesodermal patterning and blastopore lip formation. Development 122:4001–4012

Melton DA (1987) Translocation of a localized maternal mRNA to the vegetal pole of *Xenopus* oocytes. Nature 328:80–82

Mosquera L, Forristall C, Zhou Y, King ML (1993) An mRNA localized to the vegetal cortex of *Xenopus* oocytes encodes a protein with nanos-like zinc finger domain. Development 117:377–386

Mowry KL (1996) Complex formation between stage-specific oocyte factors and a *Xenopus* mRNA localization element. Proc Natl Acad Sci USA 93:14608–14613

Mowry KL, Cote CA (1999) RNA sorting in *Xenopus* oocytes and embryos. FASEB J 13:435–445

Mowry KL, Melton DA (1992) Vegetal messenger RNA localization directed by a 340-nt RNA sequence element in *Xenopus* oocytes. Science 255:991–994

Norvell A, Kelley RL, Wehr K, Schupbach T (1999) Specific isoforms of squid, a *Drosophila* hnRNP, perform distinct roles in Gurken localization during oogenesis. Genes Dev 13:864–876

Oberman F, Yisraeli JK (1995) Two non-radioactive techniques for in situ hybridization to *Xenopus* oocytes. Trends Genet 11:83–84

Oh YL, Hahm B, Kim YK, Lee HK, Lee JW, Song O, Tsukiyama-Kohara K, Kohara M, Nomoto A, Jang SK (1998) Determination of functional domains in polypyrimidine-tract-binding protein. Biochem J 331:169–175

Perez I, McAfee JG, Patton JG (1997) Multiple RRMs contribute to RNA binding specificity and affinity for polypyrimidine tract binding protein. Biochemistry 36:11881–11890

Pfaff SL, Taylor WL (1992) Characterization of a *Xenopus* oocyte factor that binds to a developmentally regulated *cis*-element in the TFIIIA gene. Dev Biol 151:306–316

Pondel M, King ML (1988) Localized maternal mRNA related to transforming growth factor ß mRNA is concentrated in a cytokeratin-enriched fraction from *Xenopus* oocytes. Proc Natl Acad Sci USA 85:7612–7616

Rebagliati MR, Weeks DL, Harvey RP, Melton DA (1985) Identification and cloning of localized maternal RNAs from *Xenopus* eggs. Cell 42:769–777

Reddy BA, Kloc M, Etkin LD (1992) The cloning and characterization of a localized maternal transcript in *Xenopus laevis* whose zygotic counterpart is detected in the CNS. Mech Dev 39:143–150

Roeder AD, Gard DL (1994) Confocal microscopy of F-actin distribution in *Xenopus* oocytes. Zygote 2:111–124

Samuels M, Deshpande G, Schedl P (1998) Activities of the sex-lethal protein in RNA binding and protein:protein interactions. Nucleic Acids Res 26:2625–2637

Schnorrer F, Bohmann K, Nusslein-Volhard C (2000) The molecular motor dynein is involved in targeting swallow and bicoid RNA to the anterior pole of *Drosophila* oocytes. Nat Cell Biol 2:185–190

Schroeder KE, Yost HJ (1996) *Xenopus* poly (A) binding protein maternal RNA is localized during oogenesis and associated with large complexes in blastula. Dev Genet 19:268–276

Schwartz SP, Aisenthal L, Elisha Z, Oberman F, Yisraeli JK (1992) A 69-kDa RNA binding protein from *Xenopus* oocytes recognizes a common motif in two vegetally localized maternal mRNAs. Proc Natl Acad Sci USA 89:11895–11899

Stennard F, Carnac G, Gurdon JB (1996) The *Xenopus* T-box gene, *Antipodean*, encodes a vegetally localised maternal mRNA and can trigger mesoderm formation. Development 122:4179–4188

Thomsen GH, Melton DA (1993) Processed Vg1 protein is an axial mesoderm inducer in *Xenopus*. Cell 74:433–441

Wallace RA, Misulovin Z, Wiley HS (1980) Growth of anuran oocytes in serum-supplemented medium. Reprod Nutr Dev 20:699–708

Wallingford JB, Rowning BA, Vogeli KM, Rothbacher U, Fraser SE, Harland RM (2000) *Dishevelled* controls cell polarity during *Xenopus* gastrulation. Nature 405:81–85

Weeks DL, Melton DA (1987) A maternal mRNA localized to the vegetal hemisphere in *Xenopus* eggs codes for a growth factor related to TGF-ß. Cell 51:861–867

Weeks DL, Bailey C, Bullock E, Dagle J, Gururajan R, Linnen J, Longo F (1995) mRNAs localized to the animal hemisphere of *Xenopus laevis* oocytes and early embryos and the proteins that they encode. In: Lipshitz HD (ed) Localized RNAs. Landes Company, Austin, pp 173–183

Wessely O, De Robertis EM (2000) The *Xenopus* homologue of Bicaudal-C is a localized maternal mRNA that can induce endoderm formation. Development 127:2053–2062

Yisraeli JK, Melton DA (1988) The maternal mRNA Vg1 is correctly localized following injection into *Xenopus* oocytes. Nature 336:592–595

Yisraeli JK, Sokol S, Melton DA (1990) A two-step model for the localization of a maternal mRNA in *Xenopus* oocytes: involvement of microtubules and microfilaments in translocation and anchoring of Vg1 mRNA. Development 108:289–298

Zhang J, King ML (1996) *Xenopus* VegT RNA is localized to the vegetal cortex during oogenesis and encodes a novel T-box transcription factor involved in mesodermal patterning. Development 122:4119–4129

Zhang Q, Yaniv K, Oberman F, Wolke U, Git A, Fromer M, Taylor WL, Meyer D, Standart N, Raz E, Yisraeli JK (1999) Vg1 RBP intracellular distribution and evolutionarily conserved expression at multiple stages during development. Mech Dev 88:101–106

Zhou Y, King ML (1996a) Localization of Xcat-2 RNA, a putative germ plasm component, to the mitochondrial cloud in *Xenopus* stage I oocytes. Development 122:2947–2953

Zhou Y, King ML (1996b) RNA transport to the vegetal cortex of *Xenopus* oocytes. Dev Biol 179:173–183

Local Protein Synthesis in Invertebrate Axons: From Dogma to Dilemma

J. van Minnen[1] and N.I. Syed[2]

1 Introduction

In the past decade, it has become increasingly apparent that both axonal and dendritic domains of neurons have a natural propensity to synthesize proteins locally, i.e., independently of their cell bodies. A combined cell biological, molecular and neurobiological approach has now provided unequivocal evidence which serves to demonstrate not only the presence of various mRNA transcripts in the extrasomal regions, but also local protein synthesis. The identity and functional significance of various mRNAs and their translated products in axons and dendrites is also being revealed. Moreover, evidence is accumulating to suggest that the expression profiles of both mRNAs and their encoded proteins in the extrasomal regions are highly regulated by a variety of extrinsic stimuli. Because most evidence in support of axonal synthesis of proteins has primarily emanated from molluscan preparations, this review will thus remain mainly focused on invertebrate species such as squid, *Aplysia* and *Lymnaea*. Specifically, we will present evidence in support of the idea that both intact and isolated axons from the above molluscan species not only harbor a vast majority of mRNA species and protein synthetic machinery, but that they are also capable of de novo protein synthesis. The functional significance of local protein synthesis will also be discussed in the context of developmental and adult plasticity.

In the 1960s, pioneering work by Giuditta and coworkers led to the discovery that the squid giant axon can synthesize proteins independent of its somata (Giuditta et al. 1968). Since then, a variety of RNAs such as tRNA (Black and Lasek 1977) and rRNA (Giuditta et al. 1980, 1983) have been detected in the axons of various other molluscan species (*Lymnaea*, Van Minnen et al. 1988; Van Minnen 1994a; *Aplysia*, Landry et al. 1992). Discovery of such RNA species in these model systems, which are highly amenable for cellular and molecular analysis, sent several laboratories in search of mechanisms by which specific mRNAs and their translated products are targeted to axons and to determine the functional significance of de novo protein synthesis in axons. In addition,

[1] Graduate School of Neurosciences Amsterdam, Research Institute Neurosciences Vrije Universiteit, Faculty of Biology, De Boelelaan 1087, 1081 HV Amsterdam, The Netherlands
[2] Department of Cell Biology and Anatomy, Faculty of Medicine, University of Calgary, 3330 Hospital Drive, NW, Calgary, Alberta, Canada

Results and Problems in Cell Differentiation, Vol. 34
D. Richter (Ed.): Cell Polarity and Subcellular RNA Localization
© Springer-Verlag Berlin Heidelberg 2001

these observations questioned a previously held pragmatic view in vertebrates according to which proteins destined for extrasomal domains were to be synthesized only exclusively in the soma. The evidence for the presence of mRNAs in axons divided both neurobiologists and cell biologists into two camps. One school of thought continued to believe that the proteins required at the extrasomal domains were originally synthesized in the soma and subsequently transported to axons; whereas the other advocated a novel concept according to which the axonal demands for various proteins were to be met locally by de novo synthesis.

In mammals, although dendritic ability to synthesize protein de novo has since been well established, similar conclusive evidence (with a few notable exceptions, Eng et al. 1999) for axonal synthesis of proteins has not yet been obtained. Apparently, the arguments used to negate axonal ability of local protein synthesis in higher animals were based primarily on EM data which failed to reveal ribosomes in these extrasomal regions (Peters et al. 1970).

Both lower vertebrate and invertebrate axons, on the other hand, have not only been shown to contain ribosomes (Martin et al. 1998; Sotelo et al. 1999) but molecular biological, biochemical and imaging techniques have since been used to provide unequivocal evidence that the axons can indeed synthesize proteins de novo (Crispino et al. 1993a, b, 1997; Van Minnen et al. 1997; Spencer et al. 1998, 2000). Since most conclusive evidence in favor of local protein synthesis in axons has primarily been obtained from molluscan species, these preparations will therefore be discussed here in detail.

The use of various molluscan species such as squid, *Aplysia* and *Lymnaea* has been of considerable value in understanding many aspects of neurobiology, ranging from cellular to behavioral. These animals with simpler nervous systems have individually identifiable neurons, which are large (up to 1 mm in diameter) and readily accessible for direct cellular and molecular manipulations. Moreover, identified neurons and their axons can be extracted from the intact ganglia and maintained in cell culture. In vitro-cultured neurons not only regenerate their axonal and neuritic processes, but they also re-establish specific synapses, which are similar to those seen in vivo. Thus, networks of functionally related neurons can be reconstructed in vitro where they generate rhythmic activity in a manner similar to that seen in vivo (Syed et al. 1990). Since the giant neurons can be extracted with long segments of their axons attached, these preparations are also amenable to 'transfection' of individual axons with various mRNAs species (Van Minnen et al. 1997; Spencer et al. 1998, 2000). The giant axons from squid have allowed researchers to isolate (in the absence of contaminating glial cells), the axoplasm for biochemical and molecular biological analysis (for review, see Koenig and Giuditta 1999), which subsequently led to the identification of various RNA species. Similarly, stimulus-induced local protein synthesis at specific synapses has been shown to underlie synaptic facilitation in *Aplysia* both in vivo and in vitro (Martin et al. 1998; Casadio et al. 1999).

Taken together, the above preparations offer a unique opportunity in which the axonal ability to synthesize protein locally can be put to the test. This review will thus highlight some of the most recent findings which challenge the existing dogma and demonstrate that proteins can indeed be synthesized locally by axons. A future dilemma facing most researchers, however, is understanding the cellular and molecular mechanisms by which specific protein synthesis is regulated locally and to determine its functional significance.

2 mRNA in Axons

To gain insights into the heterogeneity of various mRNA populations in the squid giant axon, Kaplan and coworkers used an RNA-cDNA hybridization technique, which revealed a heterogeneous population of mRNA molecules consisting of at least 200 different species. These axonal mRNAs accounted for approximately 5% of the total RNA that was found in the neuronal cell bodies. The identity of a number of these transcripts was subsequently deduced by using a cDNA library that was constructed from poly(A+) mRNA from the giant axons. These identified transcripts encoded for cytoskeletal proteins such as β-tubulin, β-actin and kinesin (Kaplan et al. 1992; Gioio et al. 1994) and the enzyme enolase (Chun et al. 1995).

Whereas in the squid, the identified mRNAs predominately encode cytosolic proteins, in other well-studied molluscs such as *Aplysia* and *Lymnaea*, transcripts encoding neuropeptides have also been identified in axons (Van Minnen et al. 1988; Landry et al. 1992; Van Minnen 1994a). In general, these transcripts were present in the entire axonal compartment, including axonal terminals from which neurohormones are released into the circulatory system (see Fig. 1). Similarly, in *Aplysia*, the gene products of left upper quadrant neurons (LUQ) were detected in axonal tracts in the pericardial nerve (Landry et al. 1992).

Together, the data from three different molluscan species serve to show that a heterogeneous population of mRNAs is indeed present in the axonal compartments of invertebrate neurons and, of these, a few have been individually characterized. Further analysis of cDNA libraries in these preparations will provide additional insight into the identity and abundance of various populations of axonal transcripts.

3 Translocation of mRNA

The presence of various transcripts in the axonal compartments of invertebrate neurons raises several questions that pertain to the mechanisms by which these mRNA are differentially targeted to extrasomal regions. For instance, compared with their vertebrate counterparts, neither *cis*- or *trans*-acting factors, nor cytoskeletal elements required for the transport of transcripts, have been identified in invertebrates. However, indirect evidence suggests that intricate regulatory mechanisms are responsible for the

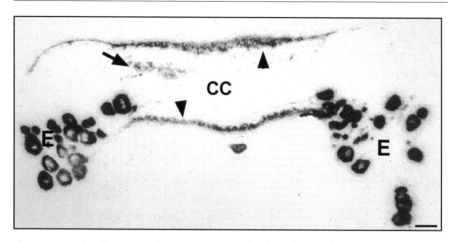

Fig. 1. mRNA localization in the axonal domain. In situ hybridization on *Lymnaea* cerebral ganglia with a ^{35}S-labeled egg-laying hormone probe revealed that the egg-laying hormone producing neurons (*E*) show a strong hybridization signal. These neurons have axons that terminate in the cerebral commissure (*CC*), where the egg-laying hormone is released into the circulation. The entire axonal compartment of these neurons is strongly labeled (*arrow* axons, *arrowheads* axon terminals). *Bar* 50 µm

targeting of selected mRNA molecules to specific axonal microdomains, such as varicosities, synapses and axon terminals. The first support for this idea comes from studies on the giant axon, where only 5% of the total somal mRNA population could be detected (see above), suggesting that a rigorous selection process must be operative in the soma for selective targeting of mRNA. Furthermore, selective translocation of transcripts has been demonstrated in both *Lymnaea* and *Aplysia*. Specifically, in *Lymnaea*, differential translocation of mRNA was observed in neurons that express neuropeptide multigene families. For instance, the LGC neurons express at least five different but related genes that encode the molluscan insulin-related peptides (MIPs; Smit et al. 1998). In situ hybridization revealed that MIP-3 transcripts were more abundant in the axonal compartment of the MIP-expressing neurons compared with MIP-1, whereas the levels of MIP-1 and -3 in the soma were more or less identical (Van Minnen 1994a). Similarly, in the ELH-expressing neurons, the ELH-1 gene is about tenfold more abundantly expressed in somata compared with the ELH-2 gene. In the axonal compartment, however, the ELH-1 gene was barely detectable in the axons, whereas ELH-2 was found to be abundantly expressed (Van Minnen 1994a). Furthermore, the extent of translocation of these neuropeptide-encoding transcripts appeared to be behaviorally regulated, since starvation (MIP-expression neurons) or egg-laying (ELH-expressing neurons) affected the amount of localized transcripts (unpubl. research). In *Aplysia*, Schacher and coworkers provided convincing evidence for selective targeting of neuropeptide (Sensorin)-encoding transcripts to distinct neuronal microdomains. Specifically, they demonstrated that cell-cell and/or synaptic contacts and electrical activity may likely regulate the branch-specific accumulation of mRNA in *Aplysia* neurites (Schacher et al. 1999).

4 Origin of RNA

In *Lymnaea* and *Aplysia*, the source of the neuropeptide-encoding mRNAs appears to be the soma as in situ hybridization revealed continuous labeling from the site of mRNA synthesis in the cell body to the axon terminals (Fig. 1). Similarly, in cultured *Aplysia* neurons, selective translocation of Sensorin mRNA has been demonstrated from the soma to specific synaptic sites and growth cones (Schacher et al. 1999).

From the above studies, one would intuitively consider the nerve cell body as the only source of mRNA in the axoplasm. In the squid giant axon, however, the axonal supply of mRNA has also been attributed to periaxonal glial cells that surround the giant axon (Rapallino et al. 1988). These experiments showed that when the isolated giant fibers (that is, without the cell bodies in the stellate ganglion) were incubated with tritiated uridine, the axoplasm incorporated considerable amounts of the radiolabel. Because the only site of transcription in this preparation was found to be the periaxonal glial cells, these studies concluded that ribonucleic acids were transported to the axoplasm from a source other than the neuronal somata. The precise mechanism underlying this glial to axonal translocation of mRNA has, however, not yet been elucidated. The ability of the glial cell to donate RNA to its (isolated) corresponding axons may, nevertheless, explain similar scenarios that have been observed in a number of vertebrate (fish, frog) and invertebrate axons (e.g. *Aplysia*, lobster). For instance, in these species when an axon is severed from its soma, the stump distal to its lesion site remains both structurally and functionally viable from months (*Aplysia*, Benbassat and Spira 1994) to a year (lobster, Parnas et al. 1991). The severed axons were shown to incorporate amino acids and were able to conduct action potentials and release neurotransmitters. The fact that these unprecedented relationships between glial cells and axons do exist was further exemplified in denucleated lobster axons. For example, Atwood and coworkers observed that when lobster axons were severed form their somata, the glia surrounding the axon invaded the axoplasm (Atwood et al. 1989). After a few days, the invading glial cells lost their cell membrane and the severed axons acquired the appearance of a multinucleated axon. These observations led the authors to postulate that this phenomenon may (partly) account for the long-term survival of transected axons. In addition to donating nuclei to the isolated axons, the surrounding periaxonal glia may also supply proteins to the axons, thus contributing to many cellular functions (see below).

5 Protein Synthesis in Axons

The first series of experiments which demonstrated that invertebrate axons are capable of protein synthesis date from as early as 1968. For instance, Giuditta and colleagues (1968) showed that the isolated squid giant axon rapidly incorporates radioactive amino acids into proteins, and that this protein synthesis can be blocked by inhibitors of ribosome-based protein synthesis, such as cycloheximide. Electrophoretic analysis further demonstrated that a wide variety of proteins are indeed synthesized locally (1968). Although it was generally accepted that the newly synthesized proteins were indeed manufactured at the periphery, the source of this de novo synthesis remained controversial. For instance, Gainer and colleagues (1977) demonstrated that the glia cells surrounding the giant axons do indeed contribute to the protein pool of axoplasm. These studies gave rise to the glia-neuron protein transfer hypothesis, according to which all newly synthesized proteins in axons are supplied by the periaxonal glial cells, and that the axon itself does not possess protein-synthesizing capabilities (Gainer et al. 1977).

Notwithstanding the fact that glial cells might contribute their proteins to axons, the data from subsequent studies showed this to be only one side of the coin. To test for axonal ability to synthesize proteins locally, Giuditta and Kaplan first set out to demonstrate that the protein synthetic machinery was indeed present and functional in the squid giant axon. These studies demonstrated that the giant axon does indeed contain elements of the eukaryotic ribosome-based protein synthesis machinery, such as elongation and initiation factors, aminoacyl-tRNA synthetases, mRNA and ribosomes (Giuditta et al. 1977, 1980, 1983, 1991; Sotelo et al. 1999). The presence of ribosomes was further confirmed by several independent techniques such as electron spectroscopic imaging, (immuno-) electron microscopy, immunocytochemistry and in situ hybridization (for further details see below). Together, these studies provided unequivocal evidence that the ribosomes are not only present in the giant axons, but also in the presynaptic ending of retinal photoreceptor neurons of the squid (Crispino et al. 1997; Martin et al. 1998). Similarly, ribosomes have also been detected in axons, growth cones and varicosities of *Lymnaea* neurons (Van Minnen 1994b; Van Minnen et al. 1997). These observations, together with the presence of tRNAs in axonal domains (Black and Lasek 1977) strongly suggest that invertebrate axons are endowed with protein synthetic capabilities. Further experimental evidence supporting axonal ability to synthesize proteins locally was derived from biochemical studies in which isolated axons were incubated with radiolabeled amino acids and the presence of radiolabeled proteins was demonstrated in the extruded axoplasm. These studies did not, however, rule out the possibility that the proteins were indeed synthesized by the axon and not its surrounding glia (local synthesis versus periaxonal glia; see above). To rule out the attributes of glial cells in local synthesis of axonal proteins, Giuditta et al. (1991) conducted a number of elegant experiments in which the isolated stellate ganglia, including the nerve con-

Fig. 2. Electrophoretic pattern of the translation product of synaptosomal polysomes purified from the squid optic lobe. *Lane 1* Synaptosomal translation products; *lane 2* after immunoabsorbtion with an anti-squid neurofilament antiserum; *lane 3* after immunoabsorbtion with a non-immune serum. Two immunopurified bands shown in *lane 2* represent 60 and 70 kDa neurofilament proteins (Grant et al. 1995). (Modified from Crispino et al. 1997)

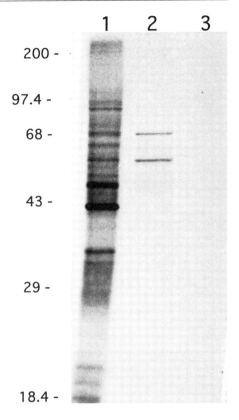

taining the giant axon, was incubated in [^{35}S]methionine. After an incubation time of 1 h, they extruded the axoplasm from the distal end of the giant axon and showed that the polyribosomes purified from the extruded axoplasm did indeed contain radiolabeled polypeptides. These studies provided direct evidence that the giant axon not only contained polyribosomes but that it was also actively involved in the synthesis of various proteins. Similar results were obtained from the synaptosomal fraction of the squid optic lobe. These data showed that the presynaptic fraction contained active polysomes, and that neurofilament proteins were synthesized in these nerve terminals (Fig. 2). Since neurofilament proteins are exclusively synthesized by neurons (Szaro et al. 1991; Way et al. 1992), these studies ruled out the possibility of glial cell involvement. The extrasomal regions in other molluscan species have also been shown to synthesize proteins in the absence of glia. For example, Kater and colleagues have demonstrated that growth cones severed from their neurites incorporate radiolabeled amino acids in cell culture. It is important to note that these cultures are absolutely devoid of glial contamination. Similarly, when *Lymnaea* neurites were severed from their somata and incubated in [^{35}S]methionine and

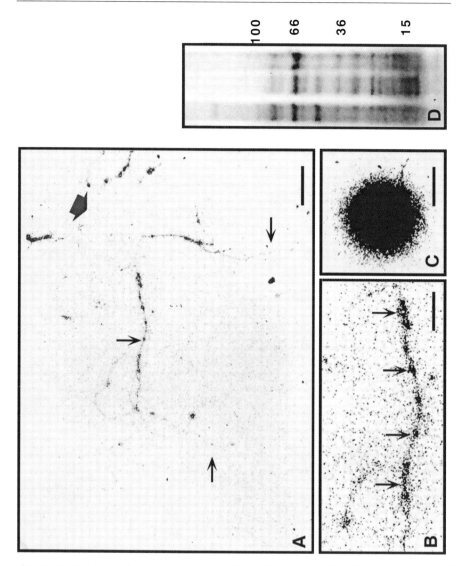

Fig. 3A–D. Protein synthesis in isolated neurites. **A** Autoradiograph of isolated neurites of *Lymnaea* pedal A neurons, incubated in ^{35}S methionine and cysteine containing 0.1 mM chloramphenicol (to inhibit mitochondrial protein synthesis) for 30 min. Labeling was observed over the neurites, indicating that these structures synthesize proteins, independent of their soma. Protein synthesis is especially prominent in varicosities (*arrows*). The *large arrow* indicates the location of the soma, before it was removed from the dish and prior to the application of the radioactive label. **B** Higher magnification of **A** highlighting protein synthesis in the varicosities of the neurites (*arrows*). **C** Autoradiograph of a soma of PeA soma, incubated in the same way as the neurites in **A**. The signal is much more intense than that of the neurites, which indicates that most protein synthesis proceeds in the somata. **D** SDS-PAGE using an 8% polyacrylamide gel. Isolated neurites from ten neurons were used for each lane. The micrograph shows that the isolated neurites produce many proteins, ranging in molecular weight from over 100 kDa to about 10 kDa. *Bar* is 50 μm in **A** and **C** and 10 μm in **B**

cysteine, newly synthesized proteins were detected by autoradiography and polyacrylamide gel electrophoresis (see Fig. 3). This neuritic protein synthesis was dependent on ribosomes, as eukaryotic ribosomal protein synthesis inhibitors, such as cyclohexamide or anisomycin effectively blocked incorporation of the radiolabeled amino acids.

6 The Identity of Protein Synthetic Machinery in Axons

To deduce the precise identity of the protein-synthesizing machinery in axons, a variety of morphological techniques (in addition to the above-described biochemical techniques) have been used to demonstrate the presence of rRNA in the squid giant axons (Giuditta et al. 1980). A first indication that axons do contain ribosomes came from experiments in which electron spectroscopic imaging was used (ESI). With this electron microscope technique, elements such as phosphorus, can be visualized because of energy loss from electrons passing through an ultrathin (10–25 nm) section. Ribosomes, rich in the element phosphorus, appear as discrete, bright particles of about 25 nm in a low-contrast microscope image (for details, see Martin et al. 1998). Using ESI, characteristic polysomal images were obtained from the squid giant axons and from the presynaptic terminals of the retinal photoreceptor neurons in the optic lobe. These ribosomes were found to be clustered in round to oval structures, called plaques (about 1 μm in size; Fig. 4). The presence of ribosomes in the giant axon and other small axons was later confirmed by immunocytochemical studies (Sotelo et al. 1999). A second approach used to demonstrate the presence of RNA, took advantage of the fluorescent dye YOYO-1. This dye has high affinity for nucleic acids and when applied to an optic lobe preparation, it generates brightly fluorescent spots in the presynaptic terminals, which have similar dimensions as those of the ESI plaques. Similar fluorescent plaques were also detected in the giant axon and these were most numerous in the postsynaptic region of the giant synapse in the stellate ganglion. The fluorescent plaques became gradually smaller and less frequent along the peripheral part of the giant axon. The ribonucleic acid content of plaques was confirmed with RNase-A treatment, which resulted in a complete disappearance of the fluorescent signal.

Ribosomes have since been demonstrated in *Lymnaea* axons both by in situ hybridization and conventional electron microscopy (Fig. 5). Using a probe for 28 S RNA, ribosomal RNA was detected in axons, growth cones and varicosities of neurons in primary culture. When the axon terminals in freshly isolated brains or primary cultured growth cones and varicosities were examined with conventional transmission electron microscopy, many electron-dense particles reminiscent of ribosomes were detected. These were often arranged in a polysome-like fashion, which is indicative of active protein synthesis. In contrast to the situation in squid, however, these ribosomes were not clustered in plaques. Other organelles involved in protein synthesis, such as rough endoplasmic reticulum (RER) and Golgi apparatus (GA) were not observed in axons of *Lymnaea* and squid, except that small patches of RER were observed in the

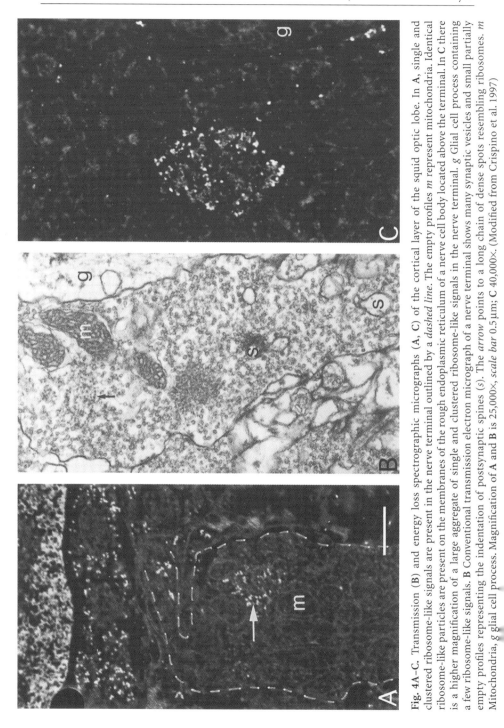

Fig. 4A–C. Transmission (**B**) and energy loss spectrographic micrographs (**A, C**) of the cortical layer of the squid optic lobe. In **A**, single and clustered ribosome-like signals are present in the nerve terminal outlined by a *dashed line*. The empty profiles *m* represent mitochondria. Identical ribosome-like particles are present on the membranes of the rough endoplasmic reticulum of a nerve cell body located above the terminal. In **C** there is a higher magnification of a large aggregate of single and clustered ribosome-like signals in the nerve terminal. *g* Glial cell process containing a few ribosome-like signals. **B** Conventional transmission electron micrograph of a nerve terminal shows many synaptic vesicles and small partially empty profiles representing the indentation of postsynaptic spines (*s*). The *arrow* points to a long chain of dense spots resembling ribosomes. *m* Mitochondria, *g* glial cell process. Magnification of **A** and **B** is 25,000×, *scale bar* 0.5 μm; **C** 40,000×. (Modified from Crispino et al. 1997)

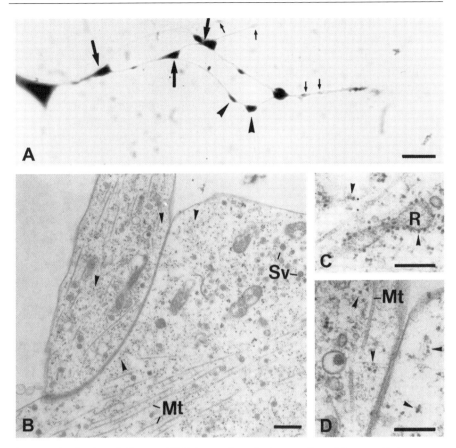

Fig. 5. A Distribution of ribosomal RNA in neurites of primary cultured *Lymnaea* PeA neurons. In situ hybridization revealed that rRNA is localized in the varicosities (*arrowheads*) and branch points (*arrows*) of PeA neurites. Note that not all of these structures show a hybridization signal (*small arrows*). *Bar* is 10 μm. **B–D** Electron micrographs of neurites (**B, D**) and soma (**C**) of *Lymnaea* neurons in primary culture. Numerous structures that have the size and appearance of ribosomes (*arrowheads*) are present in the neurites. For comparison, **C** shows a part of the soma with ribosomes (*arrowheads*) associated with the rough endoplasmic reticulum (*R*) and in the cytoplasm. *Mt* Microtubules, *Sv* synaptic vesicles. *Bar* is 0.5 μm in **B** and 100 nm in **C** and **D**

most proximal part of the giant axon (Martin et al. 1989). Together, these studies provide substantial evidence that both ribosomes and polyribosomes are indeed present in axons, though a convincing evidence for RER and GA has not yet been obtained.

7 Which Proteins Are Synthesized in Axons?

The studies described above demonstrate that axons are indeed capable of protein synthesis; they do not, however, deduce the precise identify of such proteins. Nor is it evident whether axons are selective in their ability to synthesize proteins, or if this is a function of mRNA that encodes for that specific protein. Molecular analysis of extrasomal domains has previously demonstrated that the invertebrate axons contain mRNAs that encode for three major classes of proteins; namely cytosolic proteins (e.g. cytoskeletal proteins, enzymes), luminal proteins (e.g. neuropeptides) and membrane proteins (e.g. neurotransmitter receptors). Does this serve to suggest that cognate proteins are also synthesized in the extrasomal domains? For cytosolic proteins the answer is affirmative. The axonal ability to synthesize cytoskeletal proteins has been unequivocally demonstrated in synaptosomal preparations of the squid optic lobe, where synthesis of neurofilament protein was found to depend upon polyribosomes present in these structures (Fig. 2). Although the synthesis of cytosolic proteins is known to proceed on free (poly)ribosomes present in the cytoplasm, the synthesis of luminal and receptor proteins, however, poses additional demands on the peripheral protein-synthesis machinery. The other two categories of proteins are dependent on additional organelles involving signal-recognition particles, RER and Golgi zones. These organelles are absent in the axonal territory (Peters et al. 1970; see above).

To test the axonal ability to synthesize luminal and cytosolic proteins, we used identified pedal A (PeA) cluster neurons from *Lymnaea*. We reasoned that because these neurons contain large quantities of luminal (pedal neuropeptide) transcripts in vivo, they should possess appropriate protein synthesizing machinery that would allow them to translate various exogenously expressed mRNAs. If successful, then these studies should provide compelling evidence for the translation of neuropeptide-encoding transcripts in extrasomal domains. To this end, we isolated PeA neurons, along with a long segment of their proximal axon (length up to 300 μm). After plating these neurons in culture, their corresponding axons were severed approximately one soma diameter away from the cell body, and subsequently injected with mRNA encoding a neuropeptide that is not intrinsically expressed in these PeA neurons (the egg-laying hormone, ELH, encoding mRNA). To detect translated product in the injected axon, we used an antibody against the ELH peptide. Three hours after injection, a strong fluorescent signal was detected throughout the injected portion of the axon, whereas the soma showed only background levels of immunoreactivity (Fig. 6). Although these studies firmly established the ability of neurites or axons to translate a transcript that encoded for luminal protein into its cognate protein, these experiments did not answer questions that pertained to the post-translational modifications (processing of the prohormone), routing (packaging in secretory vesicles) or the release of the newly synthesized proteins. The answers to these questions pose technical challenges as these proteins are synthesized in only minute quantities and are beyond the detectable limits of conventional biochemical approaches. Future studies, utilizing the 'sniffer' cell approach (Syed et al. 1996) may help to determine whether the ELH peptide is indeed released from the injected axons and whether this release is activity-dependent.

To investigate whether *Lymnaea* neurites are able to synthesize an integral membrane protein, we tested the axonal ability to synthesize a G-protein coupled receptor protein. In these studies, we used the conopressin receptor that was recently cloned in *Lymnaea*. Conopressin is the *Lymnaea* homologue of vasopressin, which is expressed by identified neurons in the brain (van Kesteren et al. 1996). In these experiments, we used identified visceral F cells, having first established that they do not express the conopressin receptor endogenously. These neurons were isolated from the visceral ganglia along with a large segment of their axon stump (1–2 mm). After plating the somata and its axon, both were allowed to adhere to the culture dish and the axon was further transected from the soma. mRNA encoding for the conopressin receptor was injected into a distal part of the axon (which was detached from the cell body). The injected mRNA was provided at the C-terminus, with a nucleotide sequence encoding the hemaglutinin (HA) tag. Three hours after mRNA injection, the HA tag was successfully detected in the injected axon (Fig. 7A). Since this tag was situated at the extreme C-terminal end of the receptor, these results indicate that the entire receptor protein was synthesized, rather than a truncated protein. To determine whether the axonally synthesized receptor was integrated into the outer membrane of the axon and was physiologically active, electrophysiological techniques were used. Conopressin, the ligand for the conopressin receptor, was applied exogenously (10^{-5} M), and 3–24 h after mRNA injections, electrophysiological changes in axonal excitability were monitored by intracellular recordings. The axon responded with a strong depolarization which often resulted in spiking activity, a response that was eliminated when conopressin was omitted from the perfusion pipette (Fig. 7B).

In conclusion: (1) axons from invertebrate neurons are indeed capable of de novo protein synthesis, (2) polyribosomes are present in the axonal compartments, (3) axons can synthesize cytosolic, luminal and integral membrane proteins, (4) axons are capable of functionally integrating transmembrane proteins, and (5) the axonal ability to synthesize any given protein is the function of mRNA that encodes for these proteins.

8 Functions of Local Protein Synthesis

Axons in most vertebrate species can extend over longer distances (up to several meters in some larger animals) compared with their dendritic counterparts, which have more diffused, albeit localized growth. Due to their smaller body sizes, the length of an invertebrate axon is generally shorter; among a few notable exceptions, however, is the squid giant axon which can also reach considerable lengths. Owing to the slower rates of axonal transport, most proteins destined for distantly located synaptic regions are thus highly vulnerable to degradation (the half-life of axonal cytoskeletal proteins is estimated to be only a few days; Nixon 1980). For example, proteins supplied by fast axonal transport (rate about 0.4 μm/s) to synapses located at the extreme

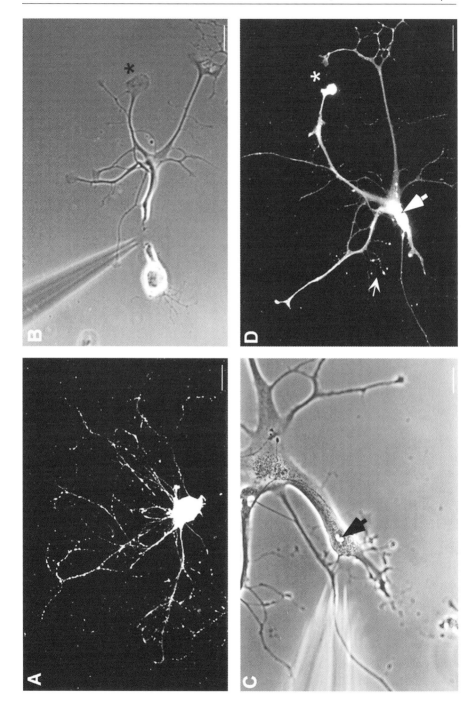

parts of the human body (about 1 m away from the soma), would take about 25 days to arrive at these distantly located areas. Even longer is the time for proteins that follow the slow axonal transport pathway, as they would travel at a rate of 1 mm/day – thus taking approximately 3 years to arrive at their destination. Although in some instances, local synthesis can be attributed to surrounding glia, this phenomenon does not explain how non-neuronal cells can supply proteins that are neuron-specific. To maintain the structural and functional integrity of long axons, it is therefore only logical to perceive that various proteins must be synthesized locally. In this manner, not only can the local demands for various proteins be met, but their expression patterns may also be regulated in a site- and protein-specific manner.

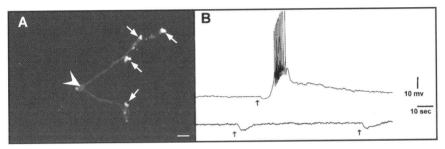

Fig. 7A,B. Functional expression of a G-protein coupled receptor in isolated axons. mRNA encoding conopressin receptor (van Kesteren et al. 1996). mRNA was injected at *arrowhead* into an axon of a visceral F neuron, isolated in culture. **A** Twenty hours following injection the conopressin receptor could be detected with immunocytochemistry, using an antiserum directed against the HA tag. *Bar* is 10 μm. **B** When conopressin (10^{-5} M) was pressure-applied onto the isolated axon (*arrows*), a strong depolarizing response leading to action potentials was detected by means of intracellular recording. This response was absent in a control-injected axon (*lower trace*). (Modified from Spencer et al. 1998)

◀──

Fig. 6A–D. Translation of egg-laying hormone mRNA in the somata (**A**) and isolated axons (**C**) of *Lymnaea* pedal A cells. **A** The egg-laying hormone encoding mRNA was injected intracellularly into the somata of sprouted pedal A cells (48 h in culture). Two to 4 h after the injections, the cells were fixed and processed for immunocytochemistry. Translation of the mRNA was detected by means of a polyclonal antiserum to egg-laying hormone (Van Minnen et al. 1988) and an FITC-conjugated secondary antiserum. The soma, neurites and varicosities showed an intense immunofluorescence. To test whether the isolated axons were able to translate mRNA encoding for egg-laying hormone, PeA cells were extracted with a large portion of their initial axon intact, plated in culture and allowed to sprout. The axon segment was severed 18-h postplating from the cell body with a micropipette (**B**) and the mRNA was injected into the axon (**C** at *arrow*). The axons were fixed 2–4 h following the injection and processed to detect egg-laying hormone immunoreactivity. Immunofluorescence was detected in the original axon (*large arrow*), varicosities (*small arrow*) and growth cones (*asterisk*). Note that following their severance from the cell body, the neurites and growth cones from the truncated axon continued to extend and were also found to be immunopositive. *Scale bar*, **A** and **B** = 100 μm, **C** = 10 μm, and **D** = 50 μm. (Modified from Van Minnen et al. 1997)

Compared with mammals, lower vertebrate and invertebrate (mollusc) axons appear to be equipped with protein synthetic machinery which is sufficient for local protein synthesis (see above). The synthesis of neurofilament proteins has been demonstrated in squid, which shows that the cytoskeletal proteins are indeed synthesized locally. Moreover, these data may also help understand how axons maintain their structural integrity in view of the metabolic break-down of structural proteins.

What other role can local protein synthesis play in peripheral domains? Important insight can perhaps be obtained from *Aplysia* (Martin et al. 1997). It is well established that the synaptic connections between neurons are constantly remodeled to meet the ever-changing behavioral demands of an animal. A similar re-arrangement of synaptic connections has also been attributed to both normal (such as learning and memory) and pathological processes (such as drug dependence, dementia and epilepsy). These changes in synaptic strength or its efficacy are site- and synapse-specific and they can be brought about by a variety of intrinsic or extrinsic stimuli.

In *Aplysia*, long-term facilitation of synaptic connections between sensory and motor neurons underlies sensitization – a form of associative learning. This form of learning consists of two distinct phases, short-term (lasting only minutes) and long-term (lasting days). At the synaptic level, these two different forms of learning are reflected in the short- and long-term facilitation of synaptic strength. Short-term facilitation requires only covalent modification (phosphorylation) of existing proteins, whereas the long- term memory requires CREB-mediated transcription and protein synthesis. Stimulating the sensory-motor synapse at least five times with 5HT (a synapse modulator), causes local protein synthesis which is necessary for long-term facilitation. This stimulation paradigm initiates CREB-mediated transcription and somal protein synthesis that is required for the expression of long-term plasticity. The newly synthesized proteins are next transported not only to the stimulated synapse but also to other unstimulated synapses. At the stimulated synapse these proteins are subsequently captured, where they persistently facilitate synaptic transmission at that particular synapse. Similarly, other synapses that were not previously subjected to 5HT application could now also be induced to undergo long-term facilitation in response to a single stimulus. This 'priming' process is also dependent on local protein synthesis. Thus, 5HT-induced local protein synthesis at synapses serves at least two different functions. Firstly, at the stimulated synapse, it serves to generate a retrograde signal that initiates CREB-mediated transcription and protein synthesis. Secondly, a single pulse of 5HT causes local protein synthesis that primes a previously unstimulated synapse for subsequent synaptic facilitation.

9 Discussion

From the above review, it becomes apparent that local protein synthesis is not only a trademark of various molluscan species, but that it also plays an important role in synaptic plasticity. It is also apparent that the expression patterns of various as yet unidentified proteins, can also be regulated in a stimulus-dependent manner. It is therefore tempting to speculate that axonal ability to synthesize protein locally may be a function of all invertebrate, if not vertebrate neurons. What is surprising, however, is that there have been no reports on axonal protein synthesis in other invertebrate species. For instance, in other invertebrate phyla, such as insects and annelids, extrasomal mRNA species have not been sited, nor has protein synthesis been demonstrated in axons. Although these invertebrate species are equally well suited for neurophysiological studies, our lack of knowledge regarding de novo protein synthesis in these animals is perplexing, and it can only be attributed to the fact that these preparations have, as yet, not been explored in this context. This will be a challenging topic for future research and the experimental models such as *Drosophila* may perhaps shed better light (compared with molluscs, which are not well suited for developmental studies) on the role of de novo local protein synthesis in neuronal development and plasticity. With the powerful genetic tools available for this animal, rapid breakthroughs regarding mechanisms regulating axonal transport and translation of mRNA can be made.

Throughout this review article we have ascribed neuronal processes the term 'axon'. Because in most invertebrates a clear distinction between axons and dendrites cannot be made, these should, in principle, be referred to as neurites. In general, invertebrates lack extreme polarization that splits them into a receptive (dendritic) and a transmittive (axonal) field. In molluscs, processes emanating from the cell body are generally termed 'axons', because these processes form presynaptic contacts on target cells, such as other neurons, muscle cells, glandular cells, etc. Moreover, incoming information is received on specialized, postsynaptic domains that are also localized at the axon. Although the molluscan axons function both in a pre- and post-synaptic capacity (which in vertebrates are distinct features of dendrites and axons, respectively), we have referred to these structures as axons.

Consistent with their dual roles (axon + dendrite), molluscan axons are capable of synthesizing both pre and postsynaptic proteins. For instance, our laboratory has shown that the mRNA encoding for neuropeptide transmitters can be translated in the isolated axons (Fig. 6), though the axonal ability to release these peptides has not yet been tested. Future studies in this model system will nevertheless allow us insights into the cellular mechanisms by which various transmitters are synthesized, packaged in vesicles and released from the extrasomal regions.

In addition to synthesizing presynaptic neurotransmitters, we have demonstrated that the isolated *Lymnaea* axons can also synthesize postsynaptic receptors from an axonally injected mRNA. The data presented in Fig. 7 clearly show that the isolated axons of VF cells not only incorporate a membrane-bound receptor into their axonal membrane, but that this receptor is also functional. A future challenge will be to elucidate the mechanisms underlying synthesis and functional integration of the conopressin receptor. Present studies do, nevertheless, provide us with both an axonal and neuronal model system in which central neurons can now be used as an expression system. This approach can also be used to alter (via selective expression of receptors and neurotransmitters) both cellular and synaptic properties of small networks of in vitro reconstructed circuits (Syed et al. 1990), and will shed further light on synaptic mechanisms underlying rhythm generation, learning and memory.

Having demonstrated that molluscan axons can synthesize proteins, and having highlighted the universal advantages of de novo protein synthesis, one now wonders why mammalian axons have been rendered incapable of protein synthesis. A careful literature review reveals that the evidence negating axonal de novo protein synthesis in mammals is itself based on negative data. Specifically, most evidence refuting the ability of mammalian axons to synthesize proteins locally comes from electron microscopy, which failed to detect ribosomes in axons (Peters et al. 1970). Furthermore, metabolic labeling studies provided evidence that axonal proteins were synthesized in the cell body and were moved towards the axonal compartment via slow and fast rates of axonal transport. Since axonal proteins do undergo metabolic decay, the question rises as to how proteins that are necessary for the functioning of the peripherally located synapses reach their destination in time, while escaping degradation. The answer may lie in experiments performed in a variety of vertebrate species (ranging from goldfish to rabbit), in which the presence of protein synthetic machinery and protein synthesis has been reported (reviewed in Koenig and Giuditta 1999). Does this mean that every mammalian axon has the ability to synthesize proteins? Most recent studies on Mauthner axons, in which ribosomes have been detected with independent morphological techniques, suggest that the axonal ability to synthesize proteins in higher vertebrates should be re-evaluated and re-examined.

A further intriguing question concerns the identity of the cellular machinery that is responsible for the production, post-translational modifications and routing of the newly synthesized proteins. Invertebrate axons synthesize a wide variety of molecules, including cytoskeletal proteins, motor proteins, secretory proteins and integral membrane proteins. The synthesis of secretory and membrane proteins and their site-specific targeting depends on rough endoplasmic reticulum (RER) and Golgi apparatus (GA). Furthermore, many proteins are post-translationally modified (processing of prohormones, glycosylation, etc). If locally synthesized proteins are indeed glycosylated, then the question arises as to which organelles are involved in this process. In the preparations (synaptosomes, axons) studied to date, the presence of polyribosomes has been demonstrated, but RER and GA were not detected. Similarly, distal dendrites of higher vertebrates do not contain a morphologically detectable RER or GA. However, for theses dendrites it was demonstrated that the cotranslational signal recognition mechanism is present (Tiedge and Brosius 1996) which may function in the post-translational modifications (i.e. glycosylation; Torre and Steward 1996). Together, these data strongly suggest that dendrites are indeed capable of synthesizing macromolecular weight proteins and of post-translational modifications that normally occur in RER and GA. The organelles held responsible for these functions are subsynaptic cisternae, membranous structures that are located beneath synapses in dendrites. These cisternae are continuous with the membranes of the endoplasmic reticulum. In addition, immunocytochemical evidence suggested that a protein residing in the GA is also present in membranous structures in dendrites. A future challenge will be to determine whether invertebrate axons possess similar organelles and whether they are the sites of macromolecular synthesis. Alternatively, in the yeast *Saccharomyces cerevisiae*, a novel mechanism for the synthesis of secretory and integral membrane proteins has recently been described. This mechanism is independent of signal recognition particles (SRPs), which are deemed necessary for the transport of protein into the lumen, or its insertion into the RER membrane. Thus proteins are synthesized on polyribosomes in the cytosol and post-translationally inserted into the RER membrane, probably by means of cytosolic heat shock protein 70 (Hsp70) and the Sec63 membrane complex (Ng et al. 1996; Schatz and Dobbersten 1996). Our unpublished immunocytochemical observations show that cultured *Lymnaea* neurites do not contain SRPs and yet they are capable of protein synthesis. At this conjecture we therefore dare to propose that protein synthesis in the isolated axons of *Lymnaea* neurons may also be achieved by mechanisms similar to those seen in yeast. Further experiments would, however, be required to test this possibility directly.

The data discussed above illustrate that invertebrate axons are capable of synthesizing many proteins. This protein synthesis is likely to serve in the general housekeeping involved in maintaining the structural and functional integrity of axons (Crispino et al. 1997), as well as in processes that underlie activity-induced synaptic plasticity (Martin et al. 1997; Casadio et al. 1999). To accommodate these processes, specific proteins have to be synthesized at the right time and at the right place. In invertebrate axons, this task appears even more daunting in view of the fact that this structure contains both pre- and postsynaptic elements, and pre- or postsynapse-specific mRNAs have to be targeted to their respective sites. This poses meticulous logistic demands on neuronal ability to route mRNAs and to regulate translation in the axonal subdomains. It can be foreseen that mRNAs encoding e.g. cytoskeletal proteins are indeed routed to different axonal domains than those encoding synaptic receptor proteins, and that their translation requires different stimuli than those of synaptic proteins. In the translocation of mRNAs, the untranslatable parts (5′and 3′ends) of the mRNAs play an important role in the regulation of transport of the mRNAs. The 3′ end, especially, has been shown to be of critical importance in regulating the translocation of many mRNA species, in animals ranging from *Drosophila* to mammals. Furthermore, several *trans*-acting proteins have been identified, especially in *Drosophila* (reviewed by St. Johnston 1995). In invertebrates, however, little is known about the mechanism underlying selective translocation of mRNA molecules. A future challenge will be the identification of *cis*- and *trans*-acting factors of invertebrate neuronal mRNAs.

   In this review we hope to have provided sufficient evidence to support our claim that the invertebrate axon, including the presynaptic endings, cannot be considered as a passive recipient of proteins derived or supplied solely from the soma and the surrounding glia cells. The axon does indeed contain a highly versatile and intricate mechanism of protein synthesis. Understanding the molecular and cellular processes underlying local protein synthesis in axons will, however, be a daunting task to achieve. The fact that a simple stimulus can exert multiple effects on local protein synthesis illustrates the complexity of regulatory processes that take place in peripheral neuronal domains. A concerted effort, utilizing various model systems and employing modern neurobiological and biochemical techniques will be required to solve the enigma underlying the mechanism by which specific proteins are synthesized and functionally regulated in the extrasomal domains.

References

Atwood HL, Dudel J, Feinstein N, Parnas I (1989) Long-term survival of decentralized axons and incorporation of satellite cells in motor neurons of rock lobsters. Neurosci Lett 101:121–126

Benbassat D, Spira ME (1994) The survival of transected axonal segments of cultured *Aplysia* neurons is prolonged by contact with intact nerve cells. Eur J Neurosci 6:1605–1614

Black MM, Lasek RJ (1977) The presence of transfer RNA in the axoplasm of the squid giant axon. J Neurobiol 8:229–237

Casadio A, Martin KC, Giustetto M, Zhu H, Chen M, Bartsch D, Bailey CH, Kandel ER (1999) A transient, neuron-wide form of CREB-mediated long-term facilitation can be stabilized at specific synapses by local protein synthesis. Cell 99:221–237

Chun JT, Gioio AE, Crispino M, Giuditta A, Kaplan BB (1995) Characterization of squid enolase mRNA: sequence analysis, tissue distribution, and axonal localization. Neurochem Res 20:923–930

Crispino M, Capano CP, Kaplan BB, Giuditta A (1993a) Neurofilament proteins are synthesized in nerve endings from squid brain. J Neurochem 61:1144–1146

Crispino M, Castigli E, Perrone Capano C, Martin R, Menichini E, Kaplan BB, Giuditta A (1993b) Protein synthesis in a synaptosomal fraction from squid brain. Mol Cell Neurosci 4:366–374

Crispino M, Kaplan BB, Martin R, Alvarez J, Chun JT, Benech JC, Giuditta A (1997) Active polysomes are present in the large presynaptic endings of the synaptosomal fraction from squid brain. J Neurosci 17:7694–7702

Eng H, Lund K, Campenot RB (1999) Synthesis of beta-tubulin, actin, and other proteins in axons of sympathetic neurons in compartmented cultures. J Neurosci 19:1–9

Gainer H, Tasaki I, Lasek R (1977) Evidence for the glia-neuron protein transfer hypothesis from intracellular perfusion studies of squid giant axons. J Cell Biol 74:524–530

Gioio AE, Chun JT, Crispino M, Perrone Capano C, Giuditta A, Kaplan BB (1994) Kinesin mRNA is present in the squid giant axon. J Neurochem 63:13–18

Giuditta A, Dettbarn WD, Brzin M (1968) Protein synthesis in the isolated giant axon of the squid. Proc Natl Acad Sci USA 59:1284–1287

Giuditta A, Metafora S, Felsani A, Del Rio A (1977) Factors for protein synthesis in the axoplasm of squid giant axons. J Neurochem 28:1391–1395

Giuditta A, Cupello A, Lazzarini G (1980) Ribosomal RNA in the axoplasm of the squid giant axon. J Neurochem 34:1757–1760

Giuditta A, Hunt T, Santella L (1983) Messenger RNA in squid axoplasm. Biol Bull 165:526

Giuditta A, Menichine E, Perrone Capano C, Langella M, Martin R, Castigli E, Kaplan BB (1991) Active polysomes in the axoplasm of the squid giant axon. J Neurosci Res 28:18–28

Grant P, Tseng D, Gould RM, Gainer H, Pant HC (1995) Expression of neurofilament proteins during development of the nervous system in the squid *Loligo pealei*. J Comp Neurol 356:311–326

Kaplan BB, Gioio AE, Perrone Capano C, Crispino M, Giuditta A (1992) Beta-actin and beta-tubulin are components of a heterogeneous mRNA population present in the squid giant axon. Mol Cell Neurosci 3:133–144

Koenig E, Giuditta A (1999) Protein-synthesizing machinery in the axon compartment. Neuroscience 89:5–15

Landry C, Crine P, DesGroseillers L (1992) Differential expression of neuropeptide gene mRNA within LUQ cells of *Aplysia californica*. J Neurobiol 23:89–101

Martin KC, Casidio A, Casadio A, Zhu HX, E YP, Rose JC, Chen M, Bailey CH, Kandel ER (1997) Synapse-specific long-term facilitation of *Aplysia* sensory to motor synapses: a function for local protein synthesis in memory storage. Cell 91:927–938

Martin R, Fritz W, Giuditta A (1989) Visualization of polyribosomes in the postsynaptic area of the squid giant synapse by electron spectroscopic imaging. J Neurocytol 18:11–18

Martin R, Vaida B, Bleher R, Crispino M, Giuditta A (1998) Protein synthesizing units in presynaptic and postsynaptic domains of squid neurons. J Cell Sci 111:3157–3166

Ng DT, Brown JD, Walter P (1996) Signal sequences specify the targeting route to the endoplasmic reticulum membrane. J Cell Biol 134:269–278

Nixon RA (1980) Protein degradation in the mouse visual system: I. Degradation of axonally transported and retinal proteins. Brain Res 200:69–83

Parnas I, Dudel J, Atwood HL (1991) Synaptic transmission in decentralized axons of rock lobster. J Neurosci 11:1309–1315

Peters S, Palay S, Webster HF (1970) The fine structure of the nervous system. Harper and Row, New York

Rapallino MV, Cupello A, Giuditta A (1988) Axoplasmic RNA species sythesized in the isolated giant axon. Neurochem Res 13:625–631

Schacher S, Wu F, Panyko JD, Sun ZY, Wang D (1999) Expression and branch-specific export of mRNA are regulated by synapse formation and interaction with specific postsynaptic targets. J Neurosci 19:6338–6347

Schatz G, Dobberstein B (1996) Common principles of protein translocation across membranes. Science 271:1519–1526

Smit AB, van Kesteren RE, Li KW, Van Minnen J, Spijker S, Van Heerikhuizen H, Geraerts WP (1998) Towards understanding the role of insulin in the brain: lessons from insulin-related signaling systems in the invertebrate brain. Prog Neurobiol 54:35–54

Sotelo JR, Kun A, Benech JC, Giuditta A, Morillas J, Benech CR (1999) Ribosomes and polyribosomes are present in the squid giant axon: an immunocytochemical study. Neuroscience 90:705–15

Spencer G, Syed NI, van Kesteren ER, Lukowiak K, Geraerts WPM, van Minnen J (2000) Synthesis and functional integration of a neurotransmitter receptor in isolated invertebrate axons. J Neurobiol 44:72–81

Spencer G, Syed NI, van Minnen J (1998) De novo protein synthesis and functional expression of a neurotransmitter receptor in isolated neurites of identified *Lymnaea* neurons. Soc Neurosci Abstr 24:1093

St Johnston D (1995) The intracellular localization of messenger RNAs. Cell 81:161–170

Syed N, Richardson P, Bulloch A (1996) Ciliary neurotrophic factor, unlike nerve growth factor, supports neurite outgrowth but not synapse formation by adult *Lymnaea* neurons. J Neurobiol 29:293–303

Syed NI, Bulloch AG, Lukowiak K (1990) In vitro reconstruction of the respiratory central pattern generator of the mollusk *Lymnaea*. Science 250:282–285

Szaro BG, Pant HC, Way J, Battey J (1991) Squid low molecular weight neurofilament proteins are a novel class of neurofilament protein. A nuclear lamin-like core and multiple distinct proteins formed by alternative RNA processing. J Biol Chem 266:15035–15041

Tiedge H, Brosius J (1996) Translational machinery in dendrites of hippocampal neurons in culture. J Neurosci 16:7171–7181

Torre ER, Steward O (1996) Protein synthesis within dendrites: glycosylation of newly synthesized proteins in dendrites of hippocampal neurons in culture. J Neurosci 16:5967–5978

Van Kesteren RE, Tensen CP, Smit AB, van Minnen J, Kolakowski LF, Meyerhof W, Richter D, van Heerikhuizen H, Vreugdenhil E, Geraerts WP (1996) Co-evolution of ligand-receptor pairs in the vasopressin/oxytocin superfamily of bioactive peptides. J Biol Chem 271:3619–3626

Van Minnen J (1994a) Axonal localization of neuropeptide-encoding mRNA in identified neurons of the snail *Lymnaea stagnalis*. Cell Tissue Res 276:155–161

Van Minnen J (1994b) RNA in the axonal domain: a new dimension in neuronal functioning? Histochem J 26:377–391

Van Minnen J, van De Haar C, Raap AK, Vreugdenhil E (1988) Localization of ovulation hormone-like neuropeptide in the central nervous system of the snail *Lymnaea stagnalis* by means of immunocytochemistry and in situ hybridization. Cell Tissue Res 251:477–484

Van Minnen J, Bergman JJ, van Kesteren ER, Smit AB, Geraerts WP, Lukowiak K, Hasan SU, Syed NI (1997) De novo protein synthesis in isolated axons of identified neurons. Neuroscience 80:1–7

Way J, Hellmich MR, Jaffe H, Szaro B, Pant HC, Gainer H, Battey J (1992) A high-molecular-weight squid neurofilament protein contains a lamin- like rod domain and a tail domain with Lys-Ser-Pro repeats. Proc Natl Acad Sci USA 89:6963–6967

Nucleocytoplasmic RNA Transport in Retroviral Replication

Harald Wodrich and Hans-Georg Kräusslich[1]

1 Introduction

Cellular gene expression is a highly ordered, stepwise process. Genetic information is transcribed in the cell nucleus into a pre-mRNA. Subsequently, this primary transcript is extensively modified on a post-transcriptional level. Intronic, non coding sequences are removed by splicing, a cap structure is added to the 5′end, and the 3′end is modified by cleavage and polyadenylation. A hallmark of all cellular gene expression is retention of the transcript in the nucleus during the entire process of modification. Only the completely processed mature RNA molecule is exported into the cytoplasm, where translation occurs.

Retroviruses integrate their genetic information into the host cell genome, where it is retained for the life of this cell and all daughter cells. The integrated retroviral genome is termed a provirus and is generally transcribed into a single primary transcript by cellular polymerase II complexes. This primary transcript encodes several open reading frames and is differentially spliced into functional mRNAs. Consequently, some retroviral mRNAs are exported into the cytoplasm as unspliced or partially spliced RNA, still containing one or several introns. Nuclear export of intron-containing RNAs is normally prohibited and several retroviruses have evolved various RNA export strategies to overcome these cellular restrictions. This RNA export needs to be quite efficient and adequately controlled, because viral replication requires large and balanced amounts of mRNAs and genomic RNA in the cytoplasm. This chapter will give a brief overview of post-transcriptional regulation of retroviral gene expression, focusing on recent developments concerning the interplay between viral and cellular factors and mainly discussing the two best-studied systems. We will not attempt to provide a complete review of this complex issue and prior work will only be summarized briefly. Furthermore, only some representative publications will be cited and the reader is referred to several excellent recent review articles for more in-depth coverage of this topic (Cullen

[1] Heinrich-Pette-Institut für experimentelle Virologie und Immunologie an der Universität Hamburg, 20251 Hamburg, Germany
Present address: H.-G. Kräusslich, Abteilung Virologie, Universität Heidelberg, Im Neuenheimer Feld 324, 69120 Heidelberg

Results and Problems in Cell Differentiation, Vol. 34
D. Richter (Ed.): Cell Polarity and Subcellular RNA Localization
© Springer-Verlag Berlin Heidelberg 2001

1998; Izaurralde and Adam 1998; Mattaj and Englmeier 1998; Pollard and Malim 1998; Stutz and Rosbash 1998; Gorlich and Kutay 1999).

2 Retroviral Replication

Retroviruses are enveloped RNA viruses. They contain two copies of a single-stranded positive-sense RNA genome of approximately 10 kb. In the infectious virion, these RNA molecules are condensed in a ribonucleoprotein complex (RNP), which is encased in a capsid shell and enveloped by a host cell-derived lipid membrane containing the viral glycoproteins. After fusion of viral and cellular membranes and injection of the RNP core into the cytoplasm, the RNA genome is reverse-transcribed into a linear double-stranded DNA by the viral enzyme reverse transcriptase, which is itself part of the RNP. The product of reverse transcription remains associated with viral proteins and this complex is termed the preintegration complex (PIC). Once the PIC reaches the nucleus, the viral DNA is inserted into the host genome by the viral enzyme integrase. The integrated provirus serves as functional expression unit. Transcriptional control elements reside in duplications of terminal sequences (long terminal repeats; LTR), with the 5' LTR serving as a promoter for polymerase II complexes and the 3' LTR providing the signal for polyadenylation.

The genomic organisation of retroviruses is basically very similar. All replication-competent retroviruses contain reading frames encoding the inner structural proteins (Gag), the replication enzymes (Pol) and the glycoproteins (Env). Complex retroviruses like HIV harbour additional open reading frames encoding regulatory proteins, which are translated from multiply spliced RNAs. In most cases, there is only one primary transcript which serves as genomic RNA and as mRNA for the viral Gag and Pol proteins. The viral glycoproteins, on the other hand, are translated from a spliced mRNA and the intron which is removed from this mRNA corresponds to the *gag-pol* coding sequences. Gag and Pol are synthesized as polyproteins on cytosolic ribosomes and are subsequently transported to the site of capsid assembly, while the glycoproteins are translated at the rough endoplasmic reticulum and transported and processed via the vesicular transport pathway. Virus release occurs in a budding process from the cell surface and is followed by extracellular conversion to the mature infectious virion. This process requires cleavage of the structural polyproteins by the viral protease (Coffin et al. 1997).

3 RNA Export in Simple versus Complex Retroviruses

All retroviruses have to overcome the same basic problem to complete productive infection: starting with only one or two primary transcripts, many different products have to be produced at defined ratios. Therefore, both unspliced and spliced RNAs need to be exported into the cytoplasm in a balanced manner and in some cases also in a temporally regulated way. The recent characterisation of retroviral RNA-export pathways has not only shed light on this important step of viral replication, but has also provided first insights into previously unknown cellular transport pathways, which are likely to play a crucial role in many other instances.

A prototype of a simple retrovirus is the D-type retrovirus Mason-Pfizer monkey virus (M-PMV), which encodes Gag, Pol and Env proteins (Fig. 1). Nucleo-cytoplasmic transport of the intron-containing *gag-pol* message (which also serves as genomic RNA) in this case is achieved by the interaction of a viral *cis*-acting sequence with cellular transport factors. This RNA element resides in the region between the *env* gene and the 3′ LTR of M-PMV (Fig. 1) and analogous elements were also identified in the related D-type retroviruses like simian retrovirus 1 and 2 (Bray et al. 1994; Zolotukhin et al. 1994). They have been termed constitutive transport elements (CTE) and will be discussed in detail later in this chapter. A *cis*-acting export element quite different from those of D-type viruses has been found in the chicken retrovirus Rous sarcoma virus (Ogert et al. 1996) and it has been suggested that all simple retroviruses contain such export sequences, which may bind to cellular export factors (see below).

Complex retroviruses like HIV-1, on the other hand, contain several additional open reading frames and gene expression in their case requires differential splicing. Besides Gag, Pol, and Env, six regulatory proteins are encoded by HIV-1 (Fig. 2) and they are derived from more than 20 mRNAs which fall into three main classes: genomic RNA, singly spliced RNAs (e.g. the env mRNA)

M-PMV genome

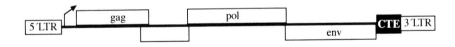

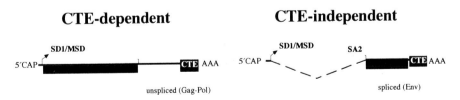

Fig. 1. Genomic organization and transcripts of the Mason-Pfizer monkey virus (M-PMV). *Top* Genome of M-PMV with open reading frames and the relative position of the constitutive transport element (CTE). *Bottom* CTE-dependent unspliced transcript encoding the Gag structural proteins (*left*) and the spliced, CTE-independent, transcript encoding the Env glycoproteins (*right*). Corresponding open reading frames and the CTE are indicated

HIV-1 genome

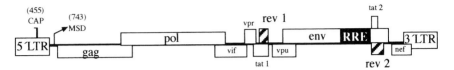

Fig. 2. Genomic organization and transcripts of the human immunodeficiency virus (HIV-1). *Top* Genomic map of HIV-1 with open reading frames. The open reading frame for Rev (*hatched boxes*) and the relative position of the RRE (*solid box*) is indicated. *Bottom* Rev-dependent transcript corresponding to the unspliced Gag-transcript and the partially spliced transcripts encoding Vif, Vpr and Vpu/Env are given on the *left*. Fully spliced, Rev-independent transcripts encoding the regulatory Rev, Tat and Nef proteins are shown on the *right*. Open reading frames and the RRE are depicted as *solid boxes*. Splice signals are shown as *arrows*. SD Splice donor, MSD major splice donor, SA splice acceptor

and multiply spliced RNAs. All incompletely spliced RNAs contain a *cis*-acting element termed Rev-responsive element (RRE). In contrast to the CTE of simple retroviruses, this sequence does not bind to cellular proteins but serves as a recognition site for the viral regulatory protein Rev, which links the corresponding RNAs to efficient cellular export pathways. This *cis-trans* regulatory system provides an additional level of control, because it allows temporal regulation of RNA export based on availability of Rev protein (summarized in Pollard and Malim 1998).

4 Regulation of HIV-1 Gene Expression: An Overview

The complex splicing pattern of HIV-1 is regulated through four different splice donor sequences and at least eight splice acceptor sites (Schwartz et al. 1990; Purcell and Martin 1993). The unspliced 9-kb primary transcript is processed into a class of partially spliced 4-kb transcripts encoding the Env protein and the regulatory factors Vif, Vpr and Vpu and into a class of completely spliced 2-kb transcripts encoding the regulatory proteins Tat, Rev and Nef (Fig. 2; Schwartz et al. 1990; Purcell and Martin 1993; Cullen 1998). Expression and cytoplasmic accumulation of all unspliced and partially spliced transcripts is dependent on the viral Rev protein, while Tat, Nef and Rev itself are expressed in a Rev-independent way (Feinberg et al. 1986; Sodroski et al. 1986). The finding that only the intron-containing transcripts of HIV-1 are Rev-dependent led to the discovery of a structured RNA element within the *env* reading frame. This element provides the Rev binding site on all intron-containing RNAs and was therefore termed the Rev responsive element (RRE). Since the RRE is part of the env coding sequence which is removed in all completely spliced RNAs, it is only present on Rev-dependent RNAs (Felber et al. 1989; Malim et al. 1989a; Malim and Cullen 1991). Rev binding to the RRE results in efficient nuclear export of the corresponding RNA (Fig. 3) via the nuclear export sequence on the Rev protein (NES; Fischer et al. 1995). Similar *cis-trans* regulatory systems have also been described for other complex retroviruses like human T-cell leukemia viruses (HTLV-I and II), Maedi-Visna virus and others (Tiley et al. 1990; Bogerd et al. 1991; Grassmann et al. 1991).

4.1 How Is Nuclear Retention of Rev-Dependent RNAs Determined?

Export of unspliced and partially spliced RNAs requires that splicing of the primary transcript is incomplete and a pool of RRE-carrying transcripts is available in the nucleus. This is indeed the case in HIV-1 infected cells, where a large amount of intron-containing nuclear transcripts are found, even in the absence of Rev (summarized in Pollard and Malim 1998). Normally, cellular pre-mRNAs are either immediately spliced to completion and then exported, or rapidly degraded (Izaurralde and Mattaj 1995). It has therefore been suggested that retroviral splice sites are generally inefficient and two different concepts have been put forward to account for the stable nuclear retention of intron-containing HIV-1 RNAs: (1) the presence of *cis* acting sequences within the intronic *gag*, *pol* and *env* sequences, and (2) the association of splice factors and other RNA-binding proteins with functional splice donor sequences present on the incompletely spliced RNA.

The first explanation is based on the finding that production of Rev-dependent HIV-1 proteins remains Rev-dependent when the corresponding coding sequences are inserted into various eukaryotic expression systems. Furthermore, fusing these sequences with heterologous RNAs confers Rev dependency onto the resulting transcripts. Several groups have identified *cis*-acting

Nucleus   Cytoplasm

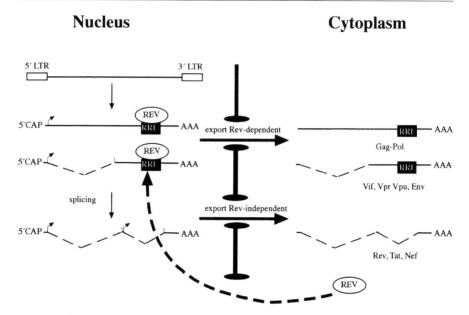

Fig. 3. Nuclear export in HIV-1 gene expression. Early in the viral life cycle the unspliced HIV-1 primary transcript is spliced in the nucleus to a class of singly and multiply spliced RNAs. Fully spliced transcripts are exported like normal cellular RNAs, resulting in the expression of the regulatory proteins Rev, Tat and Nef. Rev is reimported into the nucleus where it binds to the Rev responsive element (RRE), present only on unspliced and partially spliced transcripts. Late in viral replication, binding of Rev to the RRE results in Rev-dependent nuclear export of intron-containing viral transcripts to the cytoplasm where expression of viral structural proteins and assembly of progeny viruses occurs

RNA elements as major determinants for Rev dependency. These elements have been termed *cis*-acting repressory sequences (CRS) or inhibitory sequences (INS) and were observed in the Gag-Pol coding region (Maldarelli et al. 1991; Schwartz et al. 1992b) the Env coding region (Nasioulas et al. 1994) and in the RRE sequence itself (Brighty and Rosenberg 1994). It has been argued that they inhibit RNA export due to putative nuclear retention signals (Brighty and Rosenberg 1994; Schneider et al. 1997). Apparently, many such sequences are present within the HIV-1 coding regions, because various subregions of *gag*, *pol* and *env* conferred Rev dependency. An extensive mutagenesis approach has led to Rev-independent HIV-1 *gag* (Schwartz et al. 1992a), *gag-pol* (Schneider et al. 1997) and *env* genes (Haas et al. 1996). To achieve this, synthetic genes have been constructed where virtually all codons were altered while retaining their coding potential. CRS/INS elements appear to have no significant homology to each other and no specific binding factors, which could explain their mode of action on a molecular level, have been identified (Schneider et al. 1997).

The observation that placing the RRE into the intron of an mRNA conferred Rev dependency only when inefficient splice sites were present led to an alternative model. These experiments showed rapid and complete splicing as well as nuclear export independent of Rev, when optimal splice signals were present. However, making the splice donor inefficient led to accumulation of unspliced transcripts and their Rev-dependent nuclear export. In contrast, Rev-independent export was restored when no splice sites were present on the RNA (Chang and Sharp 1989). Based on these results, it may be suggested that the initial interaction of splice signals with the splicing machinery (also termed commitment factors) is critical to maintain a high level of intron-containing RNAs within the nucleus. In this case, absence of splice signals would lead to escape from commitment factors and nuclear retention, while Rev could override nuclear sequestration of incompletely spliced RNAs or lead to removal of splicing factors, thus making the transcripts suitable for export (Chang and Sharp 1989; Legrain and Rosbash 1989; summarized in Pollard and Malim 1998). Since all incompletely spliced RNAs by definition contain functional splice donor sequences and all retroviral splice sites are likely to be inefficient, this model can explain Rev dependency of incompletely spliced HIV-1 RNAs. Furthermore, both models may be reconciled if CRS/INS sequences correspond (at least in part) to cryptic splice signals. It should be noted, however, that experiments concerning Rev dependency are generally based on artificial expression systems and no reasoning analysis of the influence of splice sites or CRS/INS sequences in a proviral context has been published to date.

4.2 The HIV-1 Rev-RRE System

HIV-1 Rev (regulator of expression of virion proteins) is a phosphoprotein of 116 amino acids with a predominant nuclear localization and specific nucleolar enrichment under steady-state conditions. However, heterokaryon assays revealed that Rev continuously shuttles between the nucleus and the cytoplasm (Meyer and Malim 1994). It is this feature of Rev which explains its biological function. Within the nucleus, Rev binds to the RRE on incompletely spliced HIV-1 RNAs, several molecules of Rev multimerize to form an RNP complex and this complex is exported into the cytoplasm where the cargo is released and Rev is reimported into the nucleus. Two separate domains of the Rev protein are responsible for the various functions (Fig. 4). The amino terminal region contains an arginine-rich sequence which serves as a nuclear localization signal (NLS) and as an RNA binding domain (RBD; Malim et al. 1989b; Perkins et al. 1989; Bohnlein et al. 1991; summarized in Pollard and Malim 1998). Regions flanking the NLS/RBD domain are needed for multimerization of Rev. The C-terminal domain of Rev was identified as the activation domain (Malim et al. 1989a; Hope et al. 1991), which mediates nuclear export (Fischer et al. 1995; Wen et al. 1995).

Functional analysis of the interaction of Rev with its target sequence revealed that the core of the RRE corresponds to a stable stem loop of 234

A) HIV-1 Rev   B) Leucin-rich NES

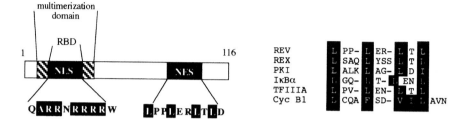

Fig. 4A,B. The HIV-1 Rev protein. **A** The Rev protein of human immunodeficiency virus harbours domains necessary for nuclear localization (*NLS*), multimerization (*hatched box*) and nuclear export (*NES*). Residues compromising the basic NLS and the leucine-rich NES are given *below*. **B** Nuclear export signals (*NES*) of the leucine-rich type from the human immuno- deficiency Rev protein, the analogue Rex protein of the human T-lymphotropic virus (HTLV I), the inhibitor of transcription factor NfκB (*IκBα*), the cAMP-dependent protein kinase inhibitor (*PKI*), transcription factor IIIA (*TFIIIA*) and cyclin B1(*Cyc B1*)

nucleotides containing a high affinity Rev binding site (Tiley et al. 1992). It is thought that binding of one Rev protein to the RRE leads to cooperative binding of additional molecules and this multimerization is essential for Rev function (Malim and Cullen 1991; Bogerd and Greene 1993). The resulting higher order RNP complex is then channeled into an export pathway via the Rev activation domain (summarized in Pollard and Malim 1998). Early exper- iments indicated that certain mutations in the activation domain led to variants (e.g. Rev M10) that inhibited Rev-function in a *trans*-dominant way (Malim et al. 1989a). Furthermore, a peptide corresponding to this region specifically inhibited Rev-mediated export, suggesting that a cellular binding factor could be titrated (Fischer et al. 1995). Transfer of the relevant sequence onto a heterologous protein rendered this protein export competent (Fischer et al. 1995), leading to the conclusion that the Rev activation domain functions as a nuclear export signal (NES). A similar NES was identified for a protein kinase inhibitor (PKI; Wen et al. 1995), indicating that the corresponding export pathway normally serves for protein transport to the cytoplasm, and has been hijacked by HIV for efficient export of its RNA (summarized in Stutz and Rosbash 1998). Accordingly, inhibition of Rev-mediated export had no effect on cellular mRNA export, but only affected the cytoplasmic accumula- tion of U snRNAs and 5 S rRNAs, which may also use NES-containing trans- port receptors (Fischer et al. 1995; Stutz and Rosbash 1998).

The regulation of HIV-1 gene expression via the Rev/RRE system provides the virus with the advantage of a temporal switch in the infection cycle. In the early phase of gene expression, only completely spliced RNAs are exported into the cytoplasm, and the regulatory proteins Tat, Rev and Nef are produced. Tat leads to increased transcription and consequently to a further increase in synthesis of early gene products. Once a sufficiently high concentration of Rev has been achieved, nuclear export of incompletely spliced mRNAs begins, thus switching to the late phase of gene expression, where translation of structural proteins and assembly of progeny virus occurs. Functionally similar early/late switches are found in many other viral systems, but are generally accomplished by the use of temporally regulated promoters, while HIV has evolved a post-transcriptional control mechanism for this purpose.

4.3 CRM1 Is the Cellular Export Receptor for Rev

All nucleocytoplasmic transport is believed to go through the nuclear pore complex (NPC). The NPC is a macromolecular structure of approximately 125 MDa which forms an aqueous channel. Transport to and from the nucleus generally uses specific import and export signals, which bind to shuttling receptors of the importin (karyopherin) family. Directionality of this transport is largely governed by the small GTPase Ran and its regulators. The GTP exchange factor (GEF) for Ran is localized in the nucleus while the GTPase activating protein (GAP) is restricted to the cytoplasm. Accordingly, there is a high nuclear and a low cytoplasmic concentration of RanGTP, which stimulates the binding of cargo to export receptors and promotes the dissociation of cargo from import receptors. Nuclear transport of the complex of cargo and receptor initiates by binding to the NPC, with subsequent transfer of the complex between different nucleoporins during movement through the NPC and eventually dissociation of the complex in the other compartment. The best characterized import signal is the classical nuclear localization signal (NLS), corresponding to a short basic sequence (Dingwall and Laskey 1991) which is also found on the HIV Rev protein. This NLS binds to the adaptor importin α, which subsequently interacts with an import receptor of the importin β family (summarized in Izaurralde and Adam 1998; Mattaj and Englmeier 1998).

The receptor for the HIV-1 Rev NES was the first nuclear export receptor described. Its identification was greatly aided by the observation that the cytotoxic drug leptomycin B (LMB) inhibits Rev export as well as Rev-dependent RNA export (Wolff et al. 1997). It had been shown that LMB resistance in the yeast *Schizosaccharomyces pombe* could be mapped to the *crm1* gene and that the corresponding gene product was the cellular target of LMB (Nishi et al. 1994). Further analysis of CRM1 in mammalian cells identified it as a distantly related member of the importin β family of proteins. Subsequently, it was shown that LMB directly inhibits Rev and U snRNA export in *Xenopus* oocytes, and this can be overcome by an excess of CRM1 (Fornerod et al. 1997a). Furthermore, LMB blocks the nuclear export of microinjected NES-fusion proteins in mammalian cells as well (Fukuda et al. 1997). It is generally believed that Rev, CRM1 and Ran-GTP form a ternary complex in a cooperative fashion, as has been shown for several other substrate-transport receptor complexes (see above). This complex may form the core of the export complex with the RRE-containing viral RNA attached as cargo (Fornerod et al. 1997a; Fukuda et al. 1997; Neville et al. 1997; Askjaer et al. 1998; summarized in Stutz and Rosbash 1998).

Rev, therefore, apparently functions as a bridging factor between the viral RNA and a cellular export pathway. What is the advantage of channeling the HIV RNAs into a pathway distinct from regular mRNA export? At present, the answer can only be speculative. Conceivably, CRM1-mediated export is more efficient, less easily saturated and faster than other export pathways, and may therefore be ideally suited for the late phase of viral replication where large amounts of RNAs encoding structural proteins must be transported to the cytoplasm. CRM1 also appears to be involved in the rapid regulation of cellular processes that are controlled by nucleocytoplasmic transport. Among the increasing number of CRM1 export substrates are kinases, transcription factors and oncogene products (summarized in Gorlich and Kutay 1999).

4.4 Other Cellular Factors That Interact with the Rev Protein

Although CRM1 has been clearly identified as an export receptor for Rev, other proteins have also been implicated in Rev-dependent nuclear export. Among these is the nucleoporin-like protein Rip1p/hRip/Rab (for Rev interacting protein/Rev activation domain binding protein; Bogerd et al. 1995; Fritz et al. 1995; Stutz et al. 1995). Rip/Rab was originally identified in a yeast two hybrid screen using Rev as a bait. In later studies it was clearly shown, however, that its interaction with Rev was indirectly mediated by CRM1 (Neville et al. 1997). Another nucleoporin (Nup 214 or CAN) has been identified as the binding protein for CRM1 (Fornerod et al. 1997b), and a variant of this protein, which lacks the C-terminal NPC binding domain, can dominantly inhibit Rev-mediated nuclear RNA export (Fornerod et al. 1997b; Bogerd et al. 1998).

Another factor that appears to play a role in Rev-mediated RNA export is the eukaryotic initiation factor 5A (eIF-5A). It was originally identified as a binding factor for a peptide corresponding to the Rev activation domain (Ruhl et al. 1993). Subsequent studies indicated that *trans*-dominant negative eIF-5 A variants efficiently inhibited the nuclear export of Rev as well as HIV replication in T-lymphocytic cell lines (Bevec et al. 1996; Junker et al. 1996). Recently, it was shown that eIF-5A interacts with CRM1, accumulates at the nuclear side of the NPC, and has an LMB-sensitive shuttling capacity (Rosorius et al. 1999). While the precise role of eIF-5A in Rev-dependent nuclear export is presently not clear, it has been suggested as an accessory factor, providing specificity to the export complex. Accordingly, *trans*-dominant eIF-5A variants can only inhibit the export of Rev and Rex proteins, while other NES-containing proteins (e.g. PKI) are not affected (Elfgang et al. 1999). Both types of NES, on the other hand, are completely blocked by specific inhibitors of CRM1 (Elfgang et al. 1999).

4.5 Rev/RRE-Like Systems in Other Complex Retroviruses

Post-transcriptional regulatory systems based on *cis*-acting sequences on the viral RNA and viral *trans*-factors have also been found in other complex retroviruses. All lentiviruses that have been analysed to date contain Rev/RRE-like systems with their respective RRE structures overlapping the *env* region. These include human immunodeficiency virus type 2, simian immunodeficiency viruses, feline immunodeficiency virus, caprine-arthritis encephalitis virus, equine infectious anaemia virus and Maedi Visna virus (summarized in Coffin M. John and Harald 1997). Besides Rev/RRE, the Rex/RxRE system of the human T-cell leukemia viruses types I and II (HTLV-1/2) is the best characterized RNA export system. Like HIV Rev, HTLV Rex is required for cytoplasmic accumulation of the unspliced genomic RNA, for synthesis of Gag-Pol proteins from this RNA and for export of partially spliced mRNAs (Inoue et al. 1987; Hidaka et al. 1988). Furthermore, the domain structure of Rex is analogous to Rev with an RNA-binding domain that also serves as NLS in the N-terminal part (Bogerd et al. 1991; Grassmann et al. 1991), and an activation domain which contains an NES-sequence, in the C-terminal part. Rex also uses CRM1 as export receptor (Bogerd et al. 1998; Hakata et al. 1998) and can efficiently substitute for Rev in mediating HIV-1 gene expression (Rimsky et al. 1988), although its primary binding site on the RRE is different. Conversely, Rev could not substitute for Rex-function in HTLV-1 gene expression, most likely due to its low affinity for the Rex responsive element (summarized in Coffin et al. 1997).

The Rex responsive element (RxRE) is located within a stem-loop structure overlapping the U3 and R regions of the HTLV 3' LTR, and not in the *env* coding region as the RRE sequence. It is therefore present on all HTLV-1 transcripts independent of whether they are spliced or not and does not discriminate between intron-containing and fully spliced RNAs. Interestingly, deletion of the 5' major splice donor from the genomic RNA resulted in Rex-independent constitutive accumulation of unspliced HTLV transcripts in the cytoplasm, indicating that splicing factors play an important role in stably retaining Rex-dependent RNAs in the nucleus (Seiki et al. 1988; Black et al. 1991).

Recently, another post-transcriptional regulatory system analogous to Rev/RRE has been identified in a family of human endogenous retroviruses (HERV-K; Magin et al. 1999; Yang et al. 1999). Endogenous retroviruses are present in the genomes of all eukaryotic cells and are transmitted vertically, thus obviating the need for extracellular infectious virions. Unlike most endogenous retroviruses, some members of the HERV-K family (e.g. HERV-K10) have maintained open reading frames for Gag, Pol and Env proteins and produce retrovirus particles, at least in certain cell lines (Patience et al. 1996). As in the case of exogenous retroviruses, these proteins are synthesized from an unspliced and singly spliced RNA, respectively. In addition, HERV-K10 produces a multiply spliced RNA encoding a protein of 105 amino acids (c-ORF) with putative NLS and NES sequences (Magin et al. 1999; Yang et al. 1999). The corresponding protein binds to its RNA target sequence, termed RcRE, in the 3′ LTR of HERV-K10 (similar to the RxRE of HTLV) and promotes export of intron-containing RNAs in a CRM1-dependent manner (Magin et al. 1999; Yang et al. 1999). C-ORF can therefore be considered a bona fide member of the family of Rev-like proteins, implying that such export factors have been present in the human genome for a long time. Conceivably, similar RNA export systems may also control differential expression of certain cellular gene products, but no examples for this have been reported to date.

5 CTE: The Constitutive RNA Transport Elements of D-Type Retroviruses

The Rev/RRE system has been identified early in the short history of HIV research and has provided an elegant explanation for the regulation of gene expression in complex retroviruses. Although simple retroviruses face a similar problem in exporting their intron-containing RNAs, no specific export mechanisms had been found for this group of viruses until recently. Initial observations showed that a short sequence downstream of the *env* region of Mason-Pfizer monkey virus (M-PMV) could efficiently substitute for the Rev/RRE system of HIV-1. Inserting this sequence into a proviral HIV-1 clone conferred Rev-independent replication, although with impaired replication capacity (Bray et al. 1994). This RNA element was termed the constitutive transport element (CTE) and was also found in other D-type retroviruses (simian retrovirus 1 and 2; SRV 1/2) and in some murine endogenous retroviruses of the intracisternal A-type particle (IAP) family (Zolotukhin et al. 1994; Tabernero et al. 1997).

Mapping the element revealed that the CTE corresponds to a structured RNA segment of 173 nucleotides in the intergenic region between *env* and the 3′ LTR. Computer prediction as well as mutational analysis and chemical probing showed that it consists of a long stable RNA helix with two internal loops and a terminal loop. The integrity of the stem and terminal loop but not their primary sequences are needed for CTE function, while the primary sequence of both internal loops is of crucial functional importance (Tabernero et al. 1996; Ernst et al. 1997a,b). The sequences of the two internal loops represent a perfect mirror image of each other and both intact loops are necessary for nuclear export and expression of CTE-dependent RNAs in mammalian cells. In contrast, microinjection of CTE-containing RNAs into *Xenopus* oocytes indicated that a single loop is sufficient for nuclear export in this case (Pasquinelli et al. 1997; Saavedra et al. 1997). It appears likely, however, that binding of multiple export factors is important for efficient CTE function since insertion of multiple CTE copies led to a significant increase in nuclear export, resulting in expression levels superior to the Rev/RRE system (Wodrich et al. 2000a). Binding of multiple molecules of an export factor to the CTE would be in analogy with the described multimerisation of Rev molecules on their target RRE sequence.

The presence of a CTE on the RNA had no influence on splicing of this RNA, and CTE-containing introns were efficiently exported from the nucleus of microinjected *Xenopus* oocytes (Pasquinelli et al. 1997; Saavedra et al. 1997). Interestingly, insertion of an intron downstream of the CTE into the 3′ untranslated region of a Rev- (or CTE)-dependent RNA led to a significant impairment of CTE function. Degradation of CTE-containing RNAs, but not of RRE-containing RNAs in the presence of Rev, was observed in this case and it was suggested that these RNAs may be subject to nonsense-mediated decay (Wodrich et al. 2000a), a recently identified degradation pathway (summarized in Hentze and Kulozik 1999).

The CTE-dependent cytoplasmic accumulation of intron-containing RNAs is independent of viral proteins, suggesting that cellular factors mediate CTE export, possibly binding to the conserved internal loop sequences. Microinjection experiments in *Xenopus* oocytes revealed that Rev-NES peptides cannot interfere with CTE-mediated export. Furthermore, microinjection of CTE-containing RNAs could efficiently block mRNA export, but not Rev-dependent export or export of U snRNA and 5 S rRNA (Fischer et al. 1995, 1999; Pasquinelli et al. 1997; Saavedra et al. 1997). Blocking CRM1, the cellular export receptor for Rev, also had no influence on CTE-mediated expression (Bogerd et al. 1998; Otero et al. 1998), indicating that CTE-mediated export does not follow the Rev/CRM1 pathway. It appears likely, on the other hand, that CTE-mediated export and cellular mRNA export share at least some limiting factor(s).

5.1 hTAP: A Potential Cellular Export Receptor for CTE

Several groups have reported putative CTE-binding factors which may be involved in RNA export. Recent experiments indicated, however, that the Tip associated protein (TAP), a homologue of the yeast mRNA export protein Mex67p (Segref et al. 1997), is the primary CTE export factor (Gruter et al. 1998). TAP was originally discovered as a cofactor for the Herpes virus saimiri oncoprotein Tip (Yoon et al. 1997). It was identified as a CTE-binding protein in cross-linking experiments, where binding to the wild-type CTE sequence was competed with CTE variants containing inactivating mutations in the internal loops. Furthermore, TAP was shown to stimulate CTE function upon microinjection into *Xenopus* oocytes (Gruter et al. 1998). The CTE binding domain of TAP was mapped to its 372 N-terminal amino acids, including a leucine-rich repeat motif (Fig. 5; Gruter et al. 1998; Braun et al. 1999; Kang and Cullen 1999). The target site for TAP on the CTE is indeed the internal RNA loop as expected for a CTE export factor (Kang et al. 1999). Microinjection and transfection experiments identified several transport domains on TAP (Fig. 5). The C-terminal region by itself is shuttling between nucleus and cytoplasm with partially overlapping NLS and NES sequences (Kang and Cullen 1999). The same domain has also been reported to direct TAP to the nuclear rim (Braun et al. 1999) and to directly interact with the FG-repeats of the nucleo-

huTAP

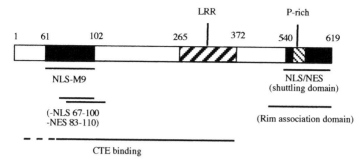

Fig. 5. The human TAP protein. Human TAP, the potential export receptor for CTE-dependent nuclear export, contains a C-terminal shuttling domain with a proline-rich motif. This domain additionally directs TAP to the nuclear Rim. CTE-binding is mediated by the N-terminus compromising the first 372 amino acids, including a leucine-arginine-rich repeat motif (LRR). A second NLS/NES with unknown function was mapped to the C-terminus (*black box*). See text for details

porin CAN/Nup214 (Katahira et al. 1999). An additional NLS was defined near the N-terminus of TAP (amino acids 61–102), which mediates nuclear import in a transportin-dependent manner (Truant et al. 1999). Another report suggested an NES (residues 83–110) partially overlapping this NLS (Bear et al. 1999). It should be noted, however, that the putative NES sequences of TAP do not share any homology with each other or with previously defined NES, and no specific export receptor for TAP has been identified so far.

CTE-dependent RNA export competes with mRNA export, indicating that TAP may normally function in mRNA export. Experiments in yeast suggest that this is indeed the case. As noted above, TAP and the yeast mRNA export factor Mex67p share a significant homology (Segref et al. 1997). Mex67p interacts with poly(A) + mRNA as well as with components of the NPC (Segref et al. 1997), and these interactions are dependent on a second factor named Mtr2p. Formation of the Mex67p/Mtr2p complex is essential for mRNA export in yeast (Santos-Rosa et al. 1998). In a later study, it was shown that TAP can be cross-linked to poly(A) + RNA in HeLa cells (Katahira et al. 1999). Furthermore, TAP interacts directly with components of the NPC and with a protein termed p15 which is related to NTF2, a Ran-GDP binding factor. The complex of TAP and p15 can functionally substitute for the Mex67p/Mtr2p complex and restore growth in an otherwise lethal yeast strain (Katahira et al. 1999). In summary, the presence of the CTE on viral RNAs appears to recruit a key factor in cellular mRNA export to incompletely spliced viral transcripts and therefore ensures their efficient nuclear export.

Other proteins have been implicated in CTE function besides TAP. Most notably, RNA helicase A (RHA) was reported to directly interact in vitro with a functional CTE but not with a non-functional variant (Tang et al. 1997). Furthermore, RHA appeared to shuttle between the nucleus and the cytoplasm, and overexpression of CTE-containing RNAs resulted in cytoplasmic accumulation of RHA and its co-localization with CTE-containing RNA (Tang et al. 1997). However, a specific RHA interaction with CTE-containing RNAs could not be confirmed by another group (Pasquinelli et al. 1997) and CTE-containing RNAs did not affect RHA localization in another study (Wodrich et al. 2000a) Given that RHA appears to be a general mRNA binding protein (Zhang et al. 1999), it is most likely associated with CTE-containing as well as with other mRNAs, but is not a CTE export factor. Two other specific CTE binding factors of ~190 and ~84 kDa have been recovered by affinity purification (Pasquinelli et al. 1997). Given their respective molecular masses, they do not correspond to the described CTE-binding proteins, but their identity has not been reported to date.

5.2 RNA Export in Other Retroviruses and Pararetroviruses

Rev/RRE and related systems have only been observed in complex retroviruses and CTEs were only found in D-type retroviruses and related elements. Other simple retroviruses face the same problem of exporting incompletely spliced RNAs, and it is presently not clear whether they all contain export elements. The only other retrovirus where a specific *cis*-acting sequence has been reported is Rous sarcoma virus (RSV), an oncogenic chicken retrovirus. The RSV genome harbours a repetitive sequence element termed direct repeat (DR) flanking the oncogene *src*. These DR elements could substitute for Rev in an HIV-1 derived expression vector in avian cells, and they were necessary for cytoplasmic accumulation of unspliced RSV transcripts. Besides promoting RNA export, they also led to enhanced stability of the RNA without affecting the splice reaction (Ogert et al. 1996; Simpson et al. 1997; Ogert and Beemon 1998). In addition, the RSV genome contains a negative regulator of splicing (NRS) which prevents the primary transcript from being efficiently spliced and leads to accumulation of unspliced RNA (McNally et al. 1991; McNally and McNally 1998; Hibbert et al. 1999). Taken together, inefficient splicing, enhanced RNA stability and facilitated nuclear export appear to work in concert to mediate balanced gene expression in this case.

Effects on RNA stability, splicing and RNA export may also be mediated by post-transcriptional control elements detected in other retroviruses and pararetroviruses. Recently, a highly structured RNA element was identified in the genome of a murine IAP, which is necessary for expression of the structural proteins of this virus and directs Rev-independent expression of HIV-1 structural proteins (Wodrich et al. 2000b). This element is clearly distinct from the CTE of D-type retroviruses, but shares homologous internal loop sequences. Besides a putative effect on RNA export, the IAP element also leads to significantly increased RNA levels, as has been reported for the RSV DR and other nonviral control elements (Ogert et al. 1996; Huang and Carmichael 1997;

Huang et al. 1999; Wodrich et al. 2000b). The pararetroviruses hepatitis B virus (HBV) and woodchuck hepatitis B virus (W-HBV) contain post transcriptional regulatory elements (PRE) which lead to enhanced stability of transcripts and stimulate the expression of a normally Rev-dependent reporter. However, the PRE does not functionally substitute for the HIV-1 Rev/RRE system in a proviral clone and does not direct expression of the Rev-dependent HIV-1 structural proteins (Schambach et al. 2000 and unpubl. observ.).

6 Summary and Conclusions

Retroviral replication is highly dependent on post-transcriptional regulation because a single primary transcript directs synthesis of many viral proteins. The identification and characterization of two post-transcriptional regulatory systems (Rev/RRE and CTE) revealed the efficient use of cellular transport pathways by retroviruses to achieve production of infectious progeny virus. The Rev/RRE system of HIV-1 consists of the viral Rev protein which binds to its target sequence on incompletely spliced RNAs and channels these into the CRM1-dependent export pathway, which is normally used for export of cellular proteins and RNAs (U snRNAs and 5 S rRNA). The CTE, on the other hand, directly recruits the cellular mRNA export receptor TAP to the viral RNA. Both systems have in common that they recruit a key player of a specific cellular export pathway and this recruitment appears to out-compete the respective cellular target molecules.

The fact that CTE can functionally substitute for Rev/RRE, yielding a replication-competent virus, indicates that very short sequence elements are sufficient for post-transcriptional control. The presence of short dominant export signals could relieve the selective pressure on the remainder of the genome to maintain a sequence that is easily exported. The resultant increase in permitted sequence space may increase the potential for immune escape, thereby providing a selective advantage for the virus. Replication of the CTE-dependent HIV-1 variant is significantly impaired compared with the wild-type virus. Considering that post-transcriptional control in the case of HIV is also used to provide a temporal switch from the early phase of regulatory protein expression to the late phase of virion production, one may suggest that the CRM1 export pathway is advantageous for the rapid delivery of large amounts of cargo (i.e. HIV RNA). This would be in accordance with its normal function because CRM1 has been shown to direct the nuclear export of cellular regulatory proteins which must be accomplished rapidly as well.

In summary, retroviruses have evolved fascinating ways to deal with their cellular environment and to make use of cellular transport pathways, allowing nuclear export of intron-containing RNAs which are normally restricted to the nucleus. Specific signals on the viral RNAs recruit key factors of cellular export, thus bypassing these restrictions and ensuring efficient viral replication.

References

Askjaer P, Jensen TH, Nilsson J, Englmeier L, Kjems J (1998) The specificity of the CRM1-Rev nuclear export signal interaction is mediated by RanGTP. J Biol Chem 273(50):33414–33422

Bear J et al (1999) Identification of novel import and export signals of human TAP, the protein that binds to the constitutive transport element of the type D retrovirus mRNAs. Mol Cell Biol 19(9):6306–6317

Bevec D et al (1996) Inhibition of HIV-1 replication in lymphocytes by mutants of the Rev cofactor eIF-5 A. Science 271(5257):1858–1860

Black AC et al (1991) Regulation of HTLV-II gene expression by Rex involves positive and negative *cis*-acting elements in the 5′ long terminal repeat. Virology 181(2):433–444

Bogerd H, Greene WC (1993) Dominant negative mutants of human T-cell leukemia virus type I Rex and human immunodeficiency virus type 1 Rev fail to multimerize in vivo. J Virol 67(5):2496–2502

Bogerd HP, Huckaby GL, Ahmed YF, Hanly SM, Greene WC (1991) The type I human T-cell leukemia virus (HTLV-I) Rex *trans*-activator binds directly to the HTLV-I Rex and the type 1 human immunodeficiency virus Rev RNA response elements. Proc Natl Acad Sci USA 88(13):5704–5708

Bogerd HP, Fridell RA, Madore S, Cullen BR (1995) Identification of a novel cellular cofactor for the Rev/Rex class of retroviral regulatory proteins. Cell 82(3):485–494

Bogerd HP, Echarri A, Ross TM, Cullen BR (1998) Inhibition of human immunodeficiency virus Rev and human T-cell leukemia virus Rex function, but not Mason-Pfizer monkey virus constitutive transport element activity, by a mutant human nucleoporin targeted to Crm1. J Virol 72(11):8627–8635

Bohnlein E, Berger J, Hauber J (1991) Functional mapping of the human immunodeficiency virus type 1 Rev RNA binding domain: new insights into the domain structure of Rev and Rex. J Virol 65(12):7051–7055

Braun IC, Rohrbach E, Schmitt C, Izaurralde E (1999) TAP binds to the constitutive transport element (CTE) through a novel RNA-binding motif that is sufficient to promote CTE-dependent RNA export from the nucleus. EMBO J 18(7):1953–1965

Bray M et al (1994) A small element from the Mason-Pfizer monkey virus genome makes human immunodeficiency virus type 1 expression and replication Rev- independent. Proc Natl Acad Sci USA 91(4):1256–1260

Brighty DW, Rosenberg M (1994) A *cis*-acting repressive sequence that overlaps the Rev-responsive element of human immunodeficiency virus type 1 regulates nuclear retention of env mRNAs independently of known splice signals. Proc Natl Acad Sci USA 91(18):8314–8318

Chang DD, Sharp PA (1989) Regulation by HIV Rev depends upon recognition of splice sites. Cell 59(5):789–795

Coffin JM, Hughes SH, Varmus HE (eds) (1997) Retroviruses. Cold Spring Harbor Laboratory Press, Cold Spring Harbor

Cullen BR (1998) Retroviruses as model systems for the study of nuclear RNA export pathways. Virology 249(2):203–210

Dingwall C, Laskey RA (1991) Nuclear targeting sequences - a consensus? (See comments.) Trends Biochem Sci 16(12):478–481

Elfgang C et al (1999) Evidence for specific nucleocytoplasmic transport pathways used by leucine-rich nuclear export signals. Proc Natl Acad Sci USA 96(11):6229–6234

Ernst RK, Bray M, Rekosh D, Hammarskjold ML (1997a) Secondary structure and mutational analysis of the Mason-Pfizer monkey virus RNA constitutive transport element. RNA 3(2):210–222

Ernst RK, Bray M, Rekosh D, Hammarskjold ML (1997b) A structured retroviral RNA element that mediates nucleocytoplasmic export of intron-containing RNA. Mol Cell Biol 17(1):135–144

Feinberg MB, Jarrett RF, Aldovini A, Gallo RC, Wong-Staal F (1986) HTLV-III expression and production involve complex regulation at the levels of splicing and translation of viral RNA. Cell 46(6):807–817

Felber BK, Hadzopoulou-Cladaras M, Cladaras C, Copeland T, Pavlakis GN (1989) Rev protein of human immunodeficiency virus type 1 affects the stability and transport of the viral mRNA. Proc Natl Acad Sci USA 86(5):1495–1499

Fischer U, Huber J, Boelens WC, Mattaj IW, Luhrmann R (1995) The HIV-1 Rev activation domain is a nuclear export signal that accesses an export pathway used by specific cellular RNAs. Cell 82(3):475–483

Fischer U et al (1999) Rev-mediated nuclear export of RNA is dominant over nuclear retention and is coupled to the Ran-GTPase cycle. Nucleic Acids Res 27(21):4128–4134

Fornerod M, Ohno M, Yoshida M, Mattaj IW (1997a) CRM1 is an export receptor for leucine-rich nuclear export signals (see comments). Cell 90(6):1051–1060

Fornerod M et al (1997b) The human homologue of yeast CRM1 is in a dynamic subcomplex with CAN/Nup214 and a novel nuclear pore component Nup88. EMBO J 16(4):807–816

Fritz CC, Zapp ML, Green MR (1995) A human nucleoporin-like protein that specifically interacts with HIV Rev. Nature 376(6540):530–533

Fukuda M et al (1997) CRM1 is responsible for intracellular transport mediated by the nuclear export signal. Nature 390(6657):308–311

Gorlich D, Kutay U (1999) Transport between the cell nucleus and the cytoplasm. Annu Rev Cell Dev Biol 15:607–660

Grassmann R et al (1991) In vitro binding of human T-cell leukemia virus rex proteins to the rex-response element of viral transcripts. J Virol 65(7):3721–3727

Gruter P et al. (1998) TAP, the human homologue of Mex67p, mediates CTE-dependent RNA export from the nucleus. Mol Cell 1(5):649–659

Haas J, Park EC, Seed B (1996) Codon usage limitation in the expression of HIV-1 envelope glycoprotein. Curr Biol 6(3):315–324

Hakata Y, Umemoto T, Matsushita S, Shida H (1998) Involvement of human CRM1 (exportin 1) in the export and multimerization of the Rex protein of human T-cell leukemia virus type 1. J Virol 72(8):6602–6607

Hentze MW, Kulozik AE (1999) A perfect message: RNA surveillance and nonsense-mediated decay. Cell 96(3):307–310

Hibbert CS, Gontarek RR, Beemon KL (1999) The role of overlapping U1 and U11 5′ splice site sequences in a negative regulator of splicing. RNA 5(3):333–343

Hidaka M, Inoue J, Yoshida M, Seiki M (1988) Post-transcriptional regulator (rex) of HTLV-1 initiates expression of viral structural proteins but suppresses expression of regulatory proteins. EMBO J 7(2):519–523

Hope TJ, Bond BL, McDonald D, Klein NP, Parslow TG (1991) Effector domains of human immunodeficiency virus type 1 Rev and human T-cell leukemia virus type I Rex are functionally interchangeable and share an essential peptide motif. J Virol 65(11):6001–6007

Huang Y, Carmichael GG (1997) The mouse histone H2a gene contains a small element that facilitates cytoplasmic accumulation of intronless gene transcripts and of unspliced HIV-1-related mRNAs. Proc Natl Acad Sci USA 94(19):10104–10109

Huang Y, Wimler KM, Carmichael GG (1999) Intronless mRNA transport elements may affect multiple steps of pre-mRNA processing. EMBO J 18(6):1642–1652

Inoue J, Yoshida M, Seiki M (1987) Transcriptional (p40x) and post-transcriptional (p27x-III) regulators are required for the expression and replication of human T-cell leukemia virus type I genes. Proc Natl Acad Sci USA 84(11):3653–3657

Izaurralde E, Adam S (1998) Transport of macromolecules between the nucleus and the cytoplasm. RNA 4(4):351–364

Izaurralde E, Mattaj IW (1995) RNA export. Cell 81(2):153–159

Junker U et al (1996) Intracellular expression of cellular eIF-5 A mutants inhibits HIV-1 replication in human T cells: a feasibility study. Hum Gene Ther 7(15):1861–1869

Kang Y, Cullen BR (1999) The human Tap protein is a nuclear mRNA export factor that contains novel RNA-binding and nucleocytoplasmic transport sequences. Genes Dev 13(9):1126–1139

Kang Y, Bogerd HP, Yang J, Cullen BR (1999) Analysis of the RNA binding specificity of the human tap protein, a constitutive transport element-specific nuclear RNA export factor. Virology 262(1):200–209

Katahira J et al. (1999) The Mex67p-mediated nuclear mRNA export pathway is conserved from yeast to human. EMBO J 18(9):2593–2609

Legrain P, Rosbash M (1989) Some *cis*- and *trans*-acting mutants for splicing target pre-mRNA to the cytoplasm. Cell 57(4):573–583

Magin C, Lower R, Lower J (1999) cOrf and RcRe, the Rev/Rex and Rre/RxRe homologues of the human endogenous retrovirus family Htdv/Herv-K. J Virol 73(11):9496–9507

Maldarelli F, Martin MA, Strebel K (1991) Identification of post-transcriptionally active inhibitory sequences in human immunodeficiency virus type 1 RNA: novel level of gene regulation. J Virol 65(11):5732–5743

Malim MH, Cullen BR (1991) HIV-1 structural gene expression requires the binding of multiple Rev monomers to the viral RRE: implications for HIV-1 latency. Cell 65(2):241–248

Malim MH, Bohnlein S, Hauber J, Cullen BR (1989a) Functional dissection of the HIV-1 Rev *trans*-activator-derivation of a *trans*-dominant repressor of Rev function. Cell 58(1):205–214

Malim MH, Hauber J, Le SY, Maizel JV, Cullen BR (1989b) The HIV-1 rev *trans*-activator acts through a structured target sequence to activate nuclear export of unspliced viral mRNA. Nature 338(6212):254–257

Mattaj IW, Englmeier L (1998) Nucleocytoplasmic transport: the soluble phase. Annu Rev Biochem 67:265–306

McNally LM, McNally MT (1998) An RNA splicing enhancer-like sequence is a component of a splicing inhibitor element from Rous sarcoma virus. Mol Cell Biol 18(6):3103–3111

McNally MT, Gontarek RR, Beemon K (1991) Characterization of Rous sarcoma virus intronic sequences that negatively regulate splicing. Virology 185(1):99–108

Meyer BE, Malim MH (1994) The HIV-1 Rev *trans*-activator shuttles between the nucleus and the cytoplasm. Genes Dev 8(13):1538–1547

Nasioulas G et al (1994) Elements distinct from human immunodeficiency virus type 1 splice sites are responsible for the Rev dependence of env mRNA. J Virol 68(5):2986–2993

Neville M, Stutz F, Lee L, Davis LI, Rosbash M (1997) The importin-beta family member Crm1p bridges the interaction between Rev and the nuclear pore complex during nuclear export. Curr Biol 7(10):767–775

Nishi K et al. (1994) Leptomycin B targets a regulatory cascade of crm1, a fission yeast nuclear protein, involved in control of higher order chromosome structure and gene expression. J Biol Chem 269(9):6320–6324

Ogert RA, Beemon KL (1998) Mutational analysis of the Rous sarcoma virus DR posttranscriptional control element. J Virol 72(4):3407–3411

Ogert RA, Lee LH, Beemon KL (1996) Avian retroviral RNA element promotes unspliced RNA accumulation in the cytoplasm. J Virol 70(6):3834–3843

Otero GC, Harris ME, Donello JE, Hope TJ (1998) Leptomycin B inhibits equine infectious anemia virus Rev and feline immunodeficiency virus rev function but not the function of the hepatitis B virus posttranscriptional regulatory element. J Virol 72(9):7593–7597

Pasquinelli AE et al (1997) The constitutive transport element (CTE) of Mason-Pfizer monkey virus (MPMV) accesses a cellular mRNA export pathway. EMBO J 16(24):7500–7510

Patience C et al (1996) Human endogenous retrovirus expression and reverse transcriptase activity in the T47D mammary carcinoma cell line. J Virol 70(4):2654–2657

Perkins A, Cochrane AW, Ruben SM, Rosen CA (1989) Structural and functional characterization of the human immunodeficiency virus rev protein. J Acquir Immune Defic Syndr 2(3):256–263

Pollard VW, Malim MH (1998) The HIV-1 Rev protein. Annu Rev Microbiol 52:491–532

Purcell DF, Martin MA (1993) Alternative splicing of human immunodeficiency virus type 1 mRNA modulates viral protein expression, replication, and infectivity. J Virol 67(11): 6365–6378

Rimsky L et al (1988) Functional replacement of the HIV-1 rev protein by the HTLV-1 rex protein. Nature 335(6192):738–740

Rosorius O et al (1999) Nuclear pore localization and nucleocytoplasmic transport of eIF-5 A: evidence for direct interaction with the export receptor CRM1. J Cell Sci 112(Pt):2369–2380

Ruhl M et al (1993) Eukaryotic initiation factor 5 A is a cellular target of the human immuno-deficiency virus type 1 Rev activation domain mediating *trans*-activation. J Cell Biol 123(6/1):1309–1320

Saavedra C, Felber B, Izaurralde E (1997) The simian retrovirus-1 constitutive transport element, unlike the HIV-1 RRE, uses factors required for cellular mRNA export. Curr Biol 7(9):619–628

Santos-Rosa H et al (1998) Nuclear mRNA export requires complex formation between Mex67p and Mtr2p at the nuclear pores. Mol Cell Biol 18(11):6826–6838

Schambach A, Wodrich H, Hildinger M, Bohne J, Krausslich H-G, Baum C (2000) Context-dependence of different modules for post-transcriptional enhancement of gene expression from retroviral vectors. Mol Therapy (in press)

Schneider R, Campbell M, Nasioulas G, Felber BK, Pavlakis GN (1997) Inactivation of the human immunodeficiency virus type 1 inhibitory elements allows Rev-independent expression of Gag and Gag/protease and particle formation. J Virol 71(7):4892–4903

Schwartz S, Felber BK, Benko DM, Fenyo EM, Pavlakis GN (1990. Cloning and functional analy-sis of multiply spliced mRNA species of human immunodeficiency virus type 1. J Virol 64(6):2519–2529

Schwartz S et al. (1992a) Mutational inactivation of an inhibitory sequence in human immuno-deficiency virus type 1 results in Rev-independent gag expression. J Virol 66(12):7176–7182

Schwartz S, Felber BK, Pavlakis GN (1992b) Distinct RNA sequences in the gag region of human immunodeficiency virus type 1 decrease RNA stability and inhibit expression in the absence of Rev protein. J Virol 66(1):150–159

Segref A et al (1997) Mex67p, a novel factor for nuclear mRNA export, binds to both poly(A)+ RNA and nuclear pores. EMBO J 16(11):3256–3271

Seiki M, Inoue J, Hidaka M, Yoshida M (1988) Two *cis*-acting elements responsible for post-transcriptional *trans*-regulation of gene expression of human T-cell leukemia virus type I. Proc Natl Acad Sci USA 85(19):7124–7128

Simpson SB, Zhang L, Craven RC, Stoltzfus CM (1997) Rous sarcoma virus direct repeat *cis* elements exert effects at several points in the virus life cycle. J Virol 71(12):9150–9156

Sodroski J et al (1986) A second post-transcriptional *trans*-activator gene required for HTLV-III replication. Nature 321(6068):412–417

Stutz F, Neville M, Rosbash M (1995) Identification of a novel nuclear pore-associated protein as a functional target of the HIV-1 Rev protein in yeast. Cell 82(3):495–506

Stutz F, Rosbash M (1998) Nuclear RNA export. Genes Dev 12(21):3303–3319

Tabernero C, Zolotukhin AS, Valentin A, Pavlakis GN, Felber BK (1996) The pos-transcriptional control element of the simian retrovirus type 1 forms an extensive RNA secondary structure necessary for its function. J Virol 70(9):5998–6011

Tabernero C et al (1997) Identification of an RNA sequence within an intracisternal-A particle element able to replace Rev-mediated post-transcriptional regulation of human immuno-deficiency virus type 1. J Virol 71(1):95–101

Tang H, Gaietta GM, Fischer WH, Ellisman MH, Wong-Staal F (1997) A cellular cofactor for the constitutive transport element of type D retrovirus. Science 276(5317):1412–1415

Tiley LS et al (1990) Visna virus encodes a post-transcriptional regulator of viral structural gene expression (published erratum appears in Proc Natl Acad Sci USA 1990, 87(23):9508). Proc Natl Acad Sci USA 87(19):7497–501

Tiley LS, Malim MH, Tewary HK, Stockley PG, Cullen BR (1992) Identification of a high-affinity RNA-binding site for the human immunodeficiency virus type 1 Rev protein (published erratum appears in Proc Natl Acad Sci USA 1992, 89(5):1997). Proc Natl Acad Sci USA 89(2):758–762

Truant R, Kang Y, Cullen BR (1999) The human tap nuclear RNA export factor contains a novel transportin-dependent nuclear localization signal that lacks nuclear export signal function. J Biol Chem 274(45):32167–32171

Wen W, Meinkoth JL, Tsien RY, Taylor SS (1995) Identification of a signal for rapid export of proteins from the nucleus. Cell 82(3):463–473

Wodrich H, Schambach A, Krausslich HG (2000a) Multiple copies of the Mason-Pfizer monkey virus constitutive RNA transport element lead to enhanced HIV-1 Gag expression in a context-dependent manner. Nucleic Acids Res 28(4):901–910

Wodrich H, Gumz E, Bohne J, Welker R, Krausslich H-G (2000b) Identification and characterization of a retroviral RNA-element in the coding region of a murine endogenous retrovirus (IAP) which can functionally replace the Rev/RRE system of HIV-1 (submitted)

Wolff B, Sanglier JJ, Wang Y (1997) Leptomycin B is an inhibitor of nuclear export: inhibition of nucleo-cytoplasmic translocation of the human immunodeficiency virus type 1 (HIV-1) Rev protein and Rev-dependent mRNA. Chem Biol 4(2):139–147

Yang J et al (1999) An ancient family of human endogenous retroviruses encodes a functional homolog of the HIV-1 Rev protein. Proc Natl Acad Sci USA 96(23):13404–13408

Yoon DW et al (1997) Tap: a novel cellular protein that interacts with tip of herpes virus saimiri and induces lymphocyte aggregation. Immunity 6(5):571–582

Zhang S, Herrmann C, Grosse F (1999) Pre-mRNA and mRNA binding of human nuclear DNA helicase II (RNA helicase A). J Cell Sci 112(7):1055–1064

Zolotukhin AS, Valentin A, Pavlakis GN, Felber BK (1994) Continuous propagation of RRE(-) and Rev(-)RRE(-) human immunodeficiency virus type 1 molecular clones containing a cis-acting element of simian retrovirus type 1 in human peripheral blood lymphocytes. J Virol 68(12):7944–7952

Subject Index

Printing (Computer to Film): Saladruck, Berlin
Binding: H. Stürtz AG, Würzburg